OCR

Physics

for AS

Chris Mee ■
Mike Crundell ■
Brian Arnold ■
Wendy Brown ■

personal tutor ›

HODDER
EDUCATION
PART OF HACHETTE LIVRE UK

Although every effort has been made to ensure that website addresses are correct at time of going to press, Hodder Education cannot be held responsible for the content of any website mentioned in this book. It is sometimes possible to find a relocated web page by typing in the address of the home page for a website in the URL window of your browser.

Hachette Livtre UK's policy is to use papers that are natural, renewable and recyclable products and made from wood grown in susainable forests. The logging and manufacturing processes are expected to conform to the environmental regulations of the country of origin.

Orders: please contact Bookpoint Ltd, 130 Milton Park, Abingdon, Oxon OX14 4SB. Telephone: (44) 01235 827720. Fax: (44) 01235 400454. Lines are open 9.00–5.00, Monday to Saturday, with a 24-hour message answering service. Visit our website at www.hoddereducation.co.uk

© Chris Mee, Mike Crundell, Brian Arnold, Wendy Brown 2008
First published in 2008 by
Hodder Education,
Part of Hachette Livre UK
338 Euston Road
London NW1 3BH

Impression number   5 4 3 2 1
Year   2013 2012 2011 2010 2009 2008

Cover photo Gustoimages/Science Photo Library
Typeset in 11/13pt Times New Roman by Charon Tec Ltd., A Macmillan Company
Printed in Italy

A catalogue record for this title is available from the British Library

ISBN: 978 0340 96779 9

Contents

Preface

This book is a revision of the AS material in *AS/A2 Physics*, amended and updated so that it is suitable for students following the OCR Physics Syllabus.

All the assessable learning outcomes are covered in this book, with some re-ordering of topics in the interests of a logical teaching order. In a few cases, the material of the book goes slightly beyond the requirements of the syllabus in order to arrive at a satisfactory termination of that topic. The assessable learning outcomes addressed in each section, or group of sections, of a chapter are listed using OCR wording, so that students can identify the exact coverage of the syllabus.

Key points, definitions and equations are highlighted in yellow panels. After each section or group of sections, there is a brief summary of the ground covered. Mathematical derivations have been kept to a minimum, but where some guidance may be helpful, this is provided in a *Maths Note* box. To achieve familiarity through practise, worked examples are provided at frequent intervals. These are followed by similar questions for students to attempt, under the heading *Now it's your turn*. In addition, there are groups of problems and questions at the end of each section or group of sections, and a selection of exam-style questions at the end of each chapter. Answers to these questions can be found at the back of the book.

A feature of the book is its Personal Tutor disc which leads the reader through selected problems on each section of the specification.

Chris Mee
Mike Crundell
Brian Arnold
Wendy Brown

1. Physical quantities and units

Physics is the study of how the world behaves and how the laws of nature operate. Accurate measurement is very important in the development of any science, and this is particularly true of physics. Throughout history, as methods of taking measurements have improved, then new ideas have developed.

In general, physicists begin by observing, measuring and collecting data. These data are then analysed to discover whether they fit into a pattern. If there is a pattern and this pattern can be used to explain other events, it becomes a theory. If the theory predicts events successfully, the theory becomes a law. The process is known as the *scientific method* (Figure 1.1).

This chapter looks at the enormous range of numbers which are met in physics, the units in which quantities are measured and how these are expressed.

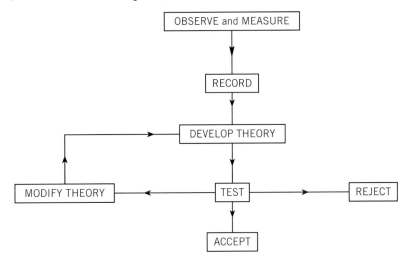

Figure 1.1 Block diagram to illustrate the scientific method

1.1 Quantities and units

At the end of Section 1.1 you should be able to:

- explain that some physical quantities consist of a numerical magnitude and a unit
- use correctly the following prefixes and their symbols to indicate decimal sub-multiples and multiples of units: pico (p), nano (n), micro (μ), milli (m), centi (c), kilo (k), mega (M), giga (G), tera (T)
- recall the following base quantities and their units: mass (kg), length (m), time (s), current (A), temperature (K), amount of substance (mol)

- express derived units as products or quotients of the base units and use these units as appropriate
- use base units to check the homogeneity of physical equations
- understand and use the conventions for labelling graph axes and table columns
- make suitable estimates of physical quantities.

Figure 1.2 Brahe (1546–1601) measured the elevations of stars; these days a modern theodolite is used for measuring angular elevation.

A **physical quantity** is a feature of something which can be measured, for example, length, weight, or time of fall. Every physical quantity has a numerical value and a unit. If someone says they have a waist measurement of 50, they could be very slim or very fat depending on whether the measurement is in centimetres or inches! Take care – it is vital to give the unit of measurement whenever a quantity is measured or written down.

Large and small quantities are usually expressed in scientific notation, i.e. as a simple number multiplied by a power of ten. For example, 0.00034 would be written as 3.4×10^{-4} and 154 000 000 as 1.54×10^8. There is far less chance of making a mistake with the number of zeros!

Figure 1.3 The elephant is large in comparison with the boy but small compared with the jumbo jet.

SI units

In very much the same way that languages have developed in various parts of the world, many different systems of measurement have evolved. Just as

Figure 1.4 The mass of this jewel could be measured in kilograms, pounds, carats, grains, etc.

languages can be translated from one to another, units of measurement can also be converted between systems. Although some conversion factors are easy to remember, some are very difficult. It is much better to have just one system of units. For this reason, scientists around the world use the **Système International (SI)** which is based on the metric system of measurement. Each quantity has just one unit and this unit can have **multiples** and **sub-multiples** to cater for larger or smaller values. The unit is given a **prefix** to denote the multiple or submultiple (Table 1.1). For example, one thousandth of a metre is known as a millimetre (mm) and 1.0 millimetre equals 1.0×10^{-3} metres (m).

prefix	symbol	multiplying factor
tera	T	10^{12}
giga	G	10^{9}
mega	M	10^{6}
kilo	k	10^{3}
centi	c	10^{-2}
milli	m	10^{-3}
micro	μ	10^{-6}
nano	n	10^{-9}
pico	p	10^{-12}

Table 1.1 The more commonly used prefixes

Beware when converting units for areas and volumes!

$$1\,mm = 10^{-3}\,m$$

Squaring both sides $\quad 1\,mm^2 = (10^{-3})^2\,m^2 = 10^{-6}\,m^2$

and $\quad 1\,mm^3 = (10^{-3})^3\,m^3 = 10^{-9}\,m^3$

Note also that $\quad 1\,cm^2 = (10^{-2})^2\,m^2 = 10^{-4}\,m^2$

and $\quad 1\,cm^3 = (10^{-2})^3\,m^3 = 10^{-6}\,m^3$

A distance of thirty metres should be written as 30 m and not 30 ms or 30 m s. The letter s is *never* included in a unit for the plural. If a space is left between two letters, the letters denote different units. So, 30 m s would mean thirty metre seconds and 30 ms means 30 milliseconds.

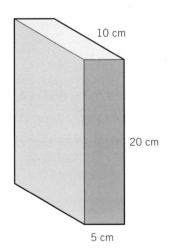

10 cm

20 cm

5 cm

Figure 1.5 This box has a volume of $1.0 \times 10^3\,cm^3$ or $1.0 \times 10^6\,mm^3$ or $1.0 \times 10^{-3}\,m^3$.

Example

Calculate the number of micrograms in 1.0 milligram.

$$1.0\,g = 1.0 \times 10^3\,mg$$

$$\text{and } 1.0\,g = 1.0 \times 10^6\,\text{micrograms } (\mu g)$$

$$\text{so, } 1.0 \times 10^3\,mg = 1.0 \times 10^6\,\mu g$$

$$\text{and } 1.0\,mg = (1.0 \times 10^6)/(1.0 \times 10^3) = \mathbf{1.0 \times 10^3\,\mu g}$$

▶▶

> ### Now it's your turn
>
> **1** Calculate the area, in cm^2, of the top of a table with sides of 1.2 m and 0.9 m.
>
> **2** Determine the number of cubic metres in one cubic kilometre.

Base units

If a quantity is to be measured accurately, the unit in which it is measured must be defined as precisely as possible.

SI is founded upon seven fundamental or **base units**.

The base quantities and the units with which they are measured are listed in Table 1.2. For completeness, the candela has been included, but this unit will not be used in this course.

Table 1.2 The base quantities

quantity	unit	symbol
mass	kilogram	kg
length	metre	m
time	second	s
electric current	ampere	A
thermodynamic temperature	kelvin	K
amount of substance	mole	mol
luminous intensity	candela	cd

Derived units

All quantities, apart from the base quantities, can be measured using **derived units**.

Derived units consist of some combination of the base units. The base units may be multiplied together or divided by one another, but never added or subtracted.

See Table 1.3 for examples of derived units. Some quantities have a named unit. For example, the unit of force is the newton, symbol N, but the newton can be expressed in terms of base units. Quantities which do not have a named unit are expressed in terms of other units. For example, acceleration is measured in metres per second per second (ms^{-2}).

Table 1.3 Some examples of derived units which may be used in this course

quantity	unit	derived unit
frequency	hertz (Hz)	s^{-1}
speed	$m\,s^{-1}$	$m\,s^{-1}$
acceleration	$m\,s^{-2}$	$m\,s^{-2}$
force	newton (N)	$kg\,m\,s^{-2}$
energy	joule (J)	$kg\,m^2\,s^{-2}$
power	watt (W)	$kg\,m^2\,s^{-3}$
electric charge	coulomb (C)	$A\,s$
potential difference	volt (V)	$kg\,m^2\,s^{-3}\,A^{-1}$
electrical resistance	ohm (Ω)	$kg\,m^2\,s^{-3}\,A^{-2}$

Example

What are the base units of speed?

Speed is defined as $\dfrac{distance}{time}$ and so the unit is $\dfrac{m}{s}$.

Division by a unit is shown using a negative index, that is s^{-1}.
The base units of speed are **$m\,s^{-1}$**.

Now it's your turn

Use the information in Tables 1.2 and 1.3 to determine the base units of the following quantities.

1 Density $\left(= \dfrac{mass}{volume} \right)$

2 Force ($= mass \times acceleration$)

Equations

It is possible to work out the total number of oranges in two bags if one bag contains four and the other five (the answer is nine!). This exercise would, of course, be nonsense if one bag contained three oranges and the other four apples. In the same way, for any equation to make sense, each term involved in the equation must have the same base units. A term in an equation is a group of numbers and symbols, and each of these terms (or groups) is added to, or subtracted from, other terms. For example, in the equation

$v = u + at$

the terms are v, u, and at.

In any equation where each term has the same base units, the equation is said to be **homogeneous** or 'balanced'.

In the example on page 5, each term has the base units m s^{-1}. If the equation is not homogeneous, then it is incorrect and is not valid. When an equation is known to be homogeneous, then the balancing of base units provides a means of finding the units of an unknown quantity.

Example

Use base units to show that the following equation is homogeneous.

work done = gain in kinetic energy + gain in gravitational potential energy

The terms in the equation are work, (gain in) kinetic energy, and (gain in) gravitational potential energy.

work done = force × distance moved

and so the base units are kg m s^{-2} × m = **kg m^2 s^{-2}**.

kinetic energy = $\frac{1}{2}$ × mass × (speed)^2

Since any pure number such as $\frac{1}{2}$ has no unit, the base units are
kg × (m s^{-1})^2 = **kg m^2 s^{-2}**.

potential energy = mass × gravitational field strength g × distance

The base units are kg × m s^{-2} × m = **kg m^2 s^{-2}**.

Conclusion: **All terms have the same base units and the equation is homogeneous.**

Now it's your turn

1 Use base units to check whether the following equations are balanced:
(a) *pressure = depth × density × gravitational field strength*
(b) *energy = mass × (speed of light)^2*

2 The thermal energy Q needed to melt a solid of mass m without any change of temperature is given by the equation

$Q = mL$

where L is a constant. Find the base units of L.

Conventions for symbols and units

You may have noticed that when symbols and units are printed, they appear in different styles of type. The symbol for a physical quantity is printed in *italic* (sloping) type, whereas its unit is in roman (upright) type. For example, velocity v is italic, but its unit m s^{-1} is roman. Of course, you will not be able to make this distinction in handwriting.

At AS level and beyond, there is a special convention for labelling columns of data in tables and graph axes. The symbol is printed first (in italic), separated by a forward slash (the printing term is a solidus) from the unit (in roman). Then the data is presented in a column, or along an axis, as pure numbers. This is illustrated in Figure 1.6, which shows a table of data and the resulting graph for the velocity v of a particle at various times t.

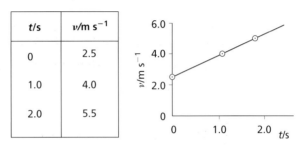

t/s	v/m s^{-1}
0	2.5
1.0	4.0
2.0	5.5

Figure 1.6 The convention for labelling tables and graphs

If you remember that a physical quantity contains a pure number and a unit, the reason for this style of presentation becomes clear. By dividing a physical quantity such as time (a number and a unit) by the appropriate unit, you are left with a pure number. It is then algebraically correct for the data in tables, and along graph axes, to appear as pure numbers.

You may also see examples in which the symbol for the physical quantity is followed by the slash, and then by a power of 10, and then the unit, for example $t/10^2$s. This means that the column of data has been divided by 100, to save repeating lots of zeros in the table. If you see a table or graph labelled $t/10^2$s and the figures 1, 2, 3 in the table column or along the graph axis, this means that the experimental data was obtained at values of t of 100 s, 200 s, 300 s.

Try to get out of the habit of heading table columns and graphs in ways such as 't in s', 't(s)' or even of recording each reading in the table as 1.0 s, 2.0 s, 3.0 s.

Order of magnitude of quantities

Figure 1.7 The ratio of the mass of the humpback whale to the mass of the mouse is about 10^4. That is minute compared to the ratio of the mass of a galaxy to the mass of a nucleus (10^{68})!

It is often useful to be able to estimate the size, or **order of magnitude**, of a quantity. Strictly speaking, the order of magnitude is the power of ten to which the number is raised. The ability to estimate is particularly important in a subject like physics where quantities have such widely different values. A *short* distance for an astrophysicist is a light-year (about 9.5×10^{15} m) whereas a *long* distance for a nuclear physicist is 6×10^{-15} m (the approximate diameter of a nucleus)! Table 1.4 gives some values of distance which may be met in this Physics course.

The ability to estimate orders of magnitude is valuable when planning and carrying out experiments or when suggesting theories. Having an idea of the expected result provides a useful check that a silly error has not been made. This is also true when using a calculator. For example, the acceleration of free fall at the Earth's surface is about 10 m s^{-2}. If a value of 9800 m s^{-2} is calculated, then this is obviously wrong and a simple error in the power of ten is likely to be the cause. Similarly, a calculation in which the cost of boiling a kettle of water is found to be several pounds, rather than a few pence, may indicate that the energy has been measured in watt hours rather than kilowatt hours.

Table 1.4 Some values of distance

	distance/m
distance from Earth to edge of observable Universe	1.4×10^{26}
diameter of a galaxy	1.2×10^{21}
distance from Earth to the Sun	1.5×10^{11}
distance from London to Paris	3.5×10^{5}
length of a car	4
diameter of a hair	5×10^{-4}
diameter of an atom	3×10^{-10}
diameter of a nucleus	6×10^{-15}

Example

It is worthwhile to remember the sizes of some common objects so that comparisons can be made. For example, a jar of peanut butter has a mass of about 500 g and a carton of orange juice has a volume of 1000 cm^3 (1 litre).

Now it's your turn

Estimate the following quantities:
(a) the mass of an orange,
(b) the mass of an adult human,
(c) the height of a room in a house,
(d) the diameter of a pencil,
(e) the volume of a small bean,
(f) the volume of a human head,
(g) the speed of a jumbo jet,
(h) the temperature of the human body.

Section 1.1 Summary

- All physical quantities have a magnitude (size) and a unit.
- The SI base units of mass, length, time, electric current, thermodynamic temperature and amount of substance are the kilogram, metre, second, ampere, kelvin and mole respectively.
- Units of all mechanical, electrical, magnetic and thermal quantities may be derived in terms of these base units.
- Physical equations must be homogeneous (balanced). Each term in an equation must have the same base units.
- The convention for printing headings in tables of data, and for labelling graph axes, is the symbol for the physical quantity (in *italic*), followed by

a forward slash, followed by the abbreviation for the unit (in roman). In handwriting, one cannot distinguish between italic and roman type.
- The order of magnitude of a number is the power of ten to which the number is raised. The order of magnitude can be used to make a check on whether a calculation gives a sensible answer.

Section 1.1 Questions

1 Write down, using scientific notation, the values of the following quantities:
    (a) 6.8 pF
    (b) 32 μC
    (c) 60 GW

2 How many electric fires, each rated at 2.5 kW, can be powered from a generator providing 2.0 MW of electric power?

3 An atom of gold, Figure 1.8, has a diameter of 0.26 nm and the diameter of its nucleus is 5.6×10^{-3} pm. Calculate the ratio of the diameter of the atom to that of the nucleus.

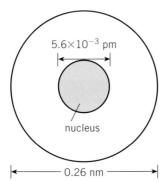

Figure 1.8 ⊢—— 0.26 nm ——⊣

4 Determine the base units of the following quantities:
    (a) energy (= *force* × *distance*)
    (b) specific heat capacity
    (*thermal energy change* = *mass* × *specific heat capacity* × *temperature change*)

5 Show that the left-hand side of the equation

    pressure + $\frac{1}{2}$ *density* × (*speed*)^2 = *constant*

    is homogeneous and find the base units of the constant on the right-hand side.

6 The period *T* of a pendulum of mass *M* is given by the expression

    $$T = 2\pi \sqrt{\frac{I}{Mgh}}$$

    where *g* is the gravitational field strength and *h* is a length.
    Determine the base units of the constant *I*.

1.2 Scalars and vectors

At the end of Section 1.2 you should be able to:
- define scalar and vector quantities and give examples of each
- represent a vector as two perpendicular components
- draw and use a vector triangle to determine the resultant of two coplanar vectors such as displacement, velocity and force
- calculate the resultant of two perpendicular vectors, such as displacement, velocity and force
- resolve a vector, such as displacement, velocity and force, into two perpendicular components
- draw and use a triangle of forces to represent the equilibrium of three forces acting at a point.

Figure 1.9 Although the athlete runs 10 km in the race, his final distance from the starting point may well be zero!

All physical quantities have a magnitude and a unit. For some quantities, magnitude and units do not give us enough information to fully describe the quantity. For example, if we are given the time for which a car travels at a certain speed, then we can calculate the distance travelled. However, we cannot find out how far the car is from its starting point unless we are told the direction of travel. In this case, the speed and direction must be specified.

A quantity which can be described fully by giving its magnitude is known as a **scalar quantity**. A **vector quantity** has magnitude and direction.

Some examples of scalar and vector quantities are given in Table 1.5.

quantity	scalar	vector
mass	✓	
weight		✓
speed	✓	
velocity		✓
force		✓
pressure	✓	
electric current		✓
temperature	✓	

Table 1.5 Some scalars and vectors

Example

A 'big wheel' at a theme park has a diameter of 14 m and people on the ride complete one revolution in 24 s. Calculate:
(a) the distance a rider moves in 3.0 minutes,
(b) the distance of the rider from the starting position.

(a) In 3.0 minutes, the rider completes $\dfrac{3.0 \times 60}{24}$ = 7.5 revolutions.

$$\text{distance travelled} = 7.5 \times \text{circumference of wheel}$$
$$= 7.5 \times 2\pi \times 7.0$$
$$= \mathbf{330\,m}$$

(b) 7.5 revolutions completed. Rider is $\frac{1}{2}$ revolution from starting point. The rider is at the opposite end of a diameter of the big wheel.
So, the distance from starting position = **14 m**.

Now it's your turn

1 State whether the following quantities are scalars or vectors:
(a) time of departure of a train,
(b) gravitational field strength,
(c) density of a liquid.

2 A student has a part-time job, earning £5 per hour. He works for 4 hours and then spends £12. Calculate and decide whether each quantity can be described as a scalar or a vector:
(a) the total of the amounts that have changed hands,
(b) the balance.

Vector representation

When you hit a tennis ball, you have to judge the direction you want it to move in, as well as how hard to hit it. The force you exert is therefore a vector quantity and cannot be represented by a number alone. One way to represent a vector is by means of an arrow. The direction of the arrow is the direction of the vector quantity. The length of the arrow, drawn to scale, represents its magnitude. This is illustrated in Figure 1.10.

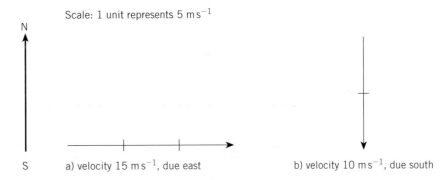

Scale: 1 unit represents 5 m s^{-1}

a) velocity 15 m s^{-1}, due east

b) velocity 10 m s^{-1}, due south

Figure 1.10 Representation of a vector quantity

Addition of vectors

The addition of two scalar quantities which have the same unit is no problem. The quantities are added using the normal rules of addition. For example, a beaker of volume 250 cm^3 and a bucket of volume 9.0 litres have a total volume of 9250 cm^3.

Adding together two vectors is more difficult because they have direction as well as magnitude. If the two vectors are in the same direction, then they can simply be added together. Two objects of weight 50 N and 40 N have a combined weight of 90 N because both weights act in the same direction (vertically downwards). Figure 1.11 shows the effect of adding two forces of magnitudes 30 N and 20 N which act in the same direction or in opposite directions. The angle between the forces is 0° when they act in the same direction and 180° when they are in opposite directions. For all other angles between the directions of the forces, the combined effect, or **resultant**, is some value between 10 N and 50 N.

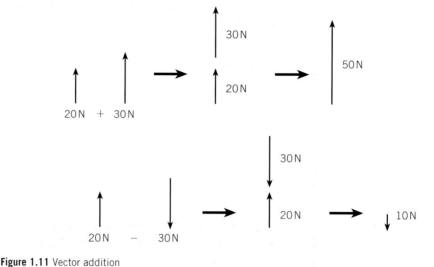

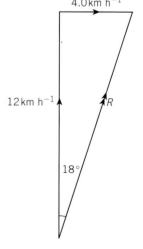

Figure 1.11 Vector addition

Figure 1.12 Vector triangles

In cases where the vectors do not act in the same or opposite directions, the resultant is found by means of a **vector triangle**. Each one of the two vectors V_1 and V_2 is represented in magnitude and direction by the side of a triangle. Note that both vectors must be in either a clockwise or an anticlockwise direction (Figure 1.12). The combined effect, or resultant R, is given in magnitude and direction by the third side of the triangle. It is important to remember that, if V_1 and V_2 are drawn clockwise, then R is anticlockwise; if V_1 and V_2 are anticlockwise, R is clockwise.

The resultant may be found by means of a scale diagram. Alternatively, having drawn a sketch of the vector triangle, the problem may be solved using trigonometry (see the Maths Note on page 17).

Example

A ship is travelling due north with a speed of $12 \, \text{km h}^{-1}$ relative to the water. There is a current in the water flowing at $4.0 \, \text{km h}^{-1}$ in an easterly direction relative to the shore. Determine the velocity of the ship relative to the shore by:
(a) scale drawing,
(b) calculation.

(a) By scale drawing (Figure 1.13):

Scale: 1 cm represents $2 \, \text{km h}^{-1}$

resultant R

The velocity relative to the shore is:

$6.3 \times 2 =$ **$12.6 \, \text{km h}^{-1}$** in a direction **18° east of north**.

Figure 1.13

(b) By calculation:

Referring to the diagram (Figure 1.14) and using Pythagoras' theorem,

$$R^2 = 12^2 + 4^2 = 160$$
$$R = \sqrt{160} = 12.6$$
$$\tan \alpha = \frac{4}{12} = 0.33$$
$$\alpha = 18.4°$$

The velocity of the ship relative to the shore is **12.6 km h⁻¹** in a direction **18.4° east of north**.

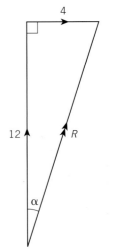

Now it's your turn

1 A swimmer who can swim in still water at a speed of 4 km h⁻¹ is swimming in a river. The river flows at a speed of 3 km h⁻¹. Calculate the speed of the swimmer relative to the river bank when she swims:
(a) downstream,
(b) upstream.

2 Draw to scale a vector triangle to determine the resultant of the two forces shown in Figure 1.15. Check your answer by calculating the resultant.

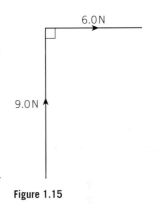

Figure 1.14

Figure 1.15

The use of a vector triangle for finding the resultant of two vectors can be demonstrated by means of a simple laboratory experiment. A weight is attached to each end of a flexible thread and the thread is then suspended over two pulleys, as shown in Figure 1.16. A third weight is attached to a point P near the centre of the thread. The string moves over the pulleys and then comes to rest. The positions of the threads are marked on a piece of paper held on a board behind the threads. This is easy to do if light from a small lamp is shone at the board. Having noted the sizes W_1 and W_2 of the weights on the ends of the thread, a vector triangle can then be drawn on the paper, as shown in Figure 1.17. The resultant of W_1 and W_2 is found to be equal in magnitude but opposite in direction to the weight W_3. If this were not so, there would be a resultant force at P and the thread and weights would move. The use of a vector triangle is justified. The three forces W_1, W_2 and W_3 are in equilibrium. The condition for the vector diagram of these forces to represent the equilibrium situation is that the three vectors should form a **closed triangle**.

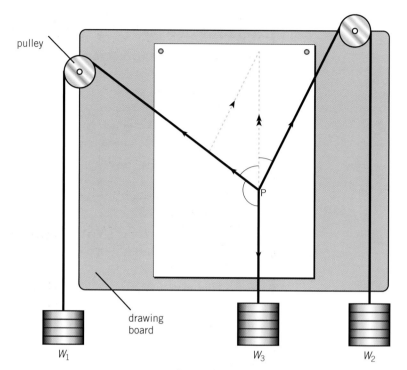

Figure 1.16 Apparatus to check the use of a vector triangle

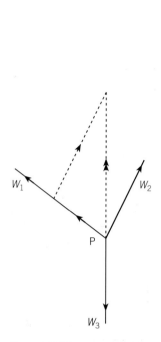

Figure 1.17 The vector triangle

We have considered only the addition of two vectors. When three or more vectors need to be added, the same principles apply, provided the vectors are **coplanar** (all in the same plane). The vector triangle then becomes a vector polygon: the resultant forms the missing side to close the polygon.

To subtract two vectors, reverse the direction (that is, change the sign) of the vector to be subtracted, and add.

Resolution of vectors

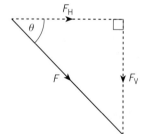

Figure 1.18 Resolving a vector into components

On pages 11–14 we saw that two vectors may be added together to produce a single resultant. This resultant behaves in the same way as the two individual vectors. It follows that a single vector may be split up, or **resolved**, into two vectors, or **components**. The combined effect of the components is the same as the original vector. In later chapters, we will see that resolution of a vector into two perpendicular components is a very useful means of solving certain types of problem.

Consider a force of magnitude F acting at an angle of θ below the horizontal (Figure 1.18). A vector triangle can be drawn with a component F_H in the horizontal direction and a component F_V acting vertically. Remembering that F, F_H and F_V form a right-angled triangle, then

$$F_H = F \cos \theta$$

and $\quad F_V = F \sin \theta$

The force F has been resolved into two perpendicular components, F_H and F_V. The example chosen is concerned with forces, but the method applies to all types of vector quantity.

Example

A glider is launched by an aircraft with a cable, as shown in Figure 1.19. At one particular moment, the tension in the cable is 620 N and the cable makes an angle of 25° with the horizontal. Calculate:

(a) the force pulling the glider horizontally,

(b) the vertical force exerted by the cable on the nose of the glider.

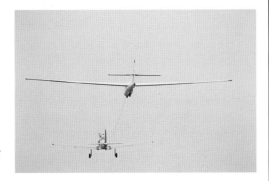

Figure 1.19

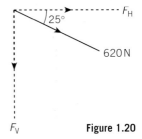

Figure 1.20

(a) horizontal component F_H = 620 cos 25
$$= \mathbf{560\,N}$$

(b) vertical component F_V = 620 sin 25 = **260 N**

Now it's your turn

1 An aircraft is travelling 35° east of north at a speed of 310 km h^{-1}. Calculate the speed of the aircraft in:
  (a) the northerly direction,
  (b) the easterly direction.

2 A cyclist is travelling down a hill at a speed of 9.2 m s^{-1}. The hillside makes an angle of 6.3° with the horizontal. Calculate, for the cyclist:
  (a) the vertical speed,
  (b) the horizontal speed.

Section 1.2 Summary

- A scalar quantity has magnitude only.
- A vector quantity has magnitude and direction.
- A vector quantity may be represented by an arrow, with the length of the arrow drawn to scale to give the magnitude.
- The combined effect of two (or more) vectors is called the resultant.
- Coplanar vectors may be added (or subtracted) using a vector diagram.
- The resultant may be found using a scale drawing of the vector diagram, or by calculation.
- A single vector may be divided into two separate components.
- The dividing of a vector into components is known as the resolution of the vector.
- In general, a vector is resolved into two components at right angles to each other.

Section 1.2 Questions

1 State whether the following quantities are scalars or vectors:
  (a) movement of the hands of a clock,
  (b) frequency of vibration,
  (c) flow of water in a pipe.

2 Speed and velocity have the same units. Explain why speed is a scalar quantity whereas velocity is a vector quantity.

3 A student states that a bag of sugar has a weight of 10 N and that this weight is a vector quantity. Discuss whether the student is correct when stating that weight is a vector.

4 Explain how an arrow may be used to represent a vector quantity.

5 Two forces are of magnitude 450 N and 240 N respectively. Determine:
  (a) the maximum magnitude of the resultant force,
  (b) the minimum magnitude of the resultant force,
  (c) the resultant force when the forces act at right angles to each other.
  Use a vector diagram and then check your result by calculation.

6 A boat can be rowed at a speed of $7.0 \, \text{km h}^{-1}$ in still water. A river flows at a constant speed of $1.5 \, \text{km h}^{-1}$. Use a scale diagram to determine the angle to the bank at which the boat must be rowed in order that the boat travels directly across the river.

7 Two forces act at a point P as shown in Figure 1.21. Draw a vector diagram, to scale, to determine the resultant force. Check your work by calculation.

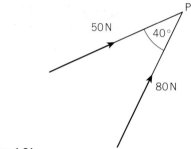

Figure 1.21

Exam-style Questions

1 (a) (i) Explain what is meant by a *base unit*.
    (ii) Give four examples of base units.
  (b) State what is meant by a *derived unit*.
  (c) (i) For any equation to be valid, it must be homogeneous. Explain what is meant by a *homogeneous* equation.
    (ii) The pressure p of an ideal gas of density ρ is given by the equation

$$p = \tfrac{1}{3}\rho\langle c^2 \rangle$$

    where $\langle c^2 \rangle$ is the mean-square-speed (i.e. it is a quantity measured as [speed]^2). Use base units to show that the equation is homogeneous.

2 (a) Determine the base units of:
    (i) work done,
    (ii) the moment of a force.
  (b) Explain why your answers to (a) mean that caution is required when the homogeneity of an equation is being tested.

3 (a) Distinguish between a *scalar* and a *vector* quantity.
  (b) A mass of weight 120 N is hung from two strings as shown in Figure 1.22.

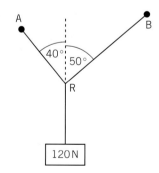

Figure 1.22

Determine, by scale drawing or by calculation, the tension in:
(i) RA,   **(ii)** RB.

(c) Use your answers in (b) to determine the horizontal component of the tension in:
(i) RA,   **(ii)** RB.
   Comment on your answer.

4 A fielder in a cricket match throws the ball to the wicket-keeper. At one moment of time, the ball has a horizontal velocity of $16\,\mathrm{m\,s^{-1}}$ and a velocity in the vertically upward direction of $8.9\,\mathrm{m\,s^{-1}}$.
(a) Determine, for the ball:
(i) its resultant speed,
(ii) the direction in which it is travelling relative to the horizontal.

(b) During the flight of the ball to the wicket-keeper, the horizontal velocity remains unchanged. The speed of the ball at the moment when the wicket-keeper catches it is $19\,\mathrm{m\,s^{-1}}$. Calculate, for the ball just as it is caught:
(i) its vertical speed,
(ii) the angle that the path of the ball makes with the horizontal.
(c) Suggest with a reason whether the ball, at the moment it is caught, is rising or falling.

Maths Note

Sine rule
For any triangle (Figure 1.23, top),

$$\frac{a}{\sin A} = \frac{b}{\sin B} = \frac{c}{\sin C}$$

Cosine rule
For any triangle,

$$a^2 = b^2 + c^2 - 2bc \cos A$$

$$b^2 = a^2 + c^2 - 2ac \cos B$$

$$c^2 = a^2 + b^2 - 2ab \cos C$$

Pythagoras' theorem
For a right-angled triangle (Figure 1.23, bottom),

$$h^2 = o^2 + a^2$$

Also for a right-angled triangle:

$$\sin \theta = \frac{o}{h}$$
$$\cos \theta = \frac{a}{h}$$
$$\tan \theta = \frac{o}{a}$$

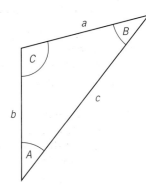
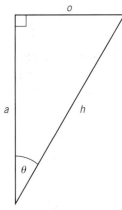

Figure 1.23

2. Kinematics

'Kinematics' is the description of how objects move. The aim of this chapter is to describe motion in terms of quantities such as position, speed, velocity and acceleration. To make matters easy, we shall talk about the motion of a particle, rather than of a real body. The simplification of dealing with a particle – a body with no size – is one that is adopted in many branches of physics.

2.1 Average speed

At the end of Sections 2.1–2.4 you should be able to:
- define displacement, instantaneous speed, average speed, velocity and acceleration
- select and use the relationships for average speed and acceleration to solve problems
- use graphical methods to represent displacement, speed, velocity and acceleration
- determine velocity from the gradient of a displacement against time graph
- determine displacement from the area under a velocity–time graph
- determine acceleration from the gradient of a velocity against time graph.

When talking about motion, we shall discuss the way in which the position of a particle varies with time. Think about a particle moving along a straight line. In a certain time, the particle will cover a certain distance. The **average speed** of the particle is defined as the distance covered divided by the time taken. Written as a word equation, this is

$$average\ speed = \frac{distance\ covered}{time\ taken}$$

The unit of speed is the metre per second (m s^{-1}).

One of the most fundamental of physical constants is the speed of light in a vacuum. It is important because it is used in the definition of the metre, and because, according to the theory of relativity, it defines an upper limit to attainable speeds. The range of speeds that you are likely to come across is enormous; some are summarised in Table 2.1.

It is important to recognise that speed has a meaning only if it is quoted relative to a fixed reference. In most cases, speeds are quoted relative to the surface of the Earth, which – although it is moving relative to the Solar

Table 2.1 Examples of speeds

	speed/m s^{-1}
light	3.0×10^8
electron around nucleus	2.2×10^6
Earth around Sun	3.0×10^4
jet airliner	2.5×10^2
typical car speed (80 km per hour)	22
sprinter	10
walking speed	1.5
snail	1×10^{-3}

System – is often taken to be fixed. Thus, when we say that a bird can fly at a certain average speed, we are relating its speed to the Earth. However, a passenger on a ferry may see that a seagull, flying parallel to the boat, appears to be practically stationary. If this is the case, the seagull's speed relative to the boat is zero. However, if the speed of the boat through the water is 8 m s^{-1}, then the speed of the seagull relative to Earth is also 8 m s^{-1}. When talking about relative speeds we must also be careful about directions. It is easy if the motions are in the same direction, as in the example of the ferry and the seagull. The addition of velocity vectors is considered in Section 1.2.

Examples

1 The radius of the Earth is 6.4×10^6 m; one revolution about its axis takes 24 hours (8.6×10^4 s). Calculate the average speed of a point on the equator relative to the centre of the Earth.

   In 24 hours, the point on the equator completes one revolution and travels a distance of $2\pi \times$ the Earth's radius, that is $2\pi \times 6.4 \times 10^6 = 4.0 \times 10^7$ m. The average speed is (distance covered)/(time taken), or $4.0 \times 10^7/8.6 \times 10^4 =$ **4.7×10^2 m s^{-1}**.

2 How far does a cyclist travel in 11 minutes if his average speed is 22 km h^{-1}?

   First convert the average speed in km h^{-1} to a value in m s^{-1}. 22 km (2.2×10^4 m) in 1 hour (3.6×10^3 s) is an average speed of 6.1 m s^{-1}. 11 minutes is 660 s. Since average speed is (distance covered)/(time taken), the distance covered is given by (average speed) × (time taken), or $6.1 \times 660 =$ **4000 m**.
   Note the importance of working in consistent units: this is why the average speed and the time were converted to m s^{-1} and s respectively.

3 A train is travelling at a speed of 25 m s^{-1} along a straight track. A boy walks along the corridor in a carriage towards the rear of the train, at a speed of 1 m s^{-1} relative to the train. What is his speed relative to Earth?

In one second, the train travels 25 m forwards along the track. In the same time the boy moves 1 m towards the rear of the train, so he has moved 24 m along the track. His speed relative to Earth is thus $25 - 1 = \mathbf{24\,m\,s^{-1}}$.

Now it's your turn

1 The speed of an electron in orbit about the nucleus of a hydrogen atom is 2.2×10^6 m s^{-1}. It takes 1.5×10^{-16} s for the electron to complete one orbit. Find the radius of the orbit.

2 The average speed of an airliner on a domestic flight is 220 m s^{-1}. How long will it take to fly between two airports on a flight-path 700 km long?

3 Two cars are travelling in the same direction on a long, straight road. The one in front has an average speed of 25 m s^{-1} relative to Earth; the other's is 31 m s^{-1}, also relative to Earth. What is the speed of the second car relative to the first when it is overtaking?

2.2 Speed and velocity

In ordinary language, there is no difference between the terms *speed* and *velocity*. However, in physics there is an important distinction between the two. **Velocity** is used to represent a vector quantity: the magnitude of how fast a particle is moving, and the direction in which it is moving. **Speed** does not have an associated direction. It is a scalar quantity (see Section 1.2).

So far, we have talked about the total distance travelled by a body. Like speed, distance is a scalar quantity, because we do not have to specify the direction in which the distance is travelled. However, in defining velocity we introduce a quantity called **displacement**. Displacement of a particle is its change of position. Consider a cyclist travelling 500 m along a straight road, and then turning round and coming back 300 m. The total distance travelled is 800 m, but the displacement is only 200 m, since the cyclist has ended up 200 m from the starting point.

The **average velocity** is defined as the displacement divided by the time taken.

$$average\ velocity = \frac{displacement}{time\ taken}$$

Because distance and displacement are different quantities, the average speed of motion will sometimes be different from the magnitude of the average velocity. If the time taken for the cyclist's trip in the example above is 120 s, the average speed is $800/120 = 6.7$ m s^{-1}, whereas the magnitude of the average velocity is $200/120 = 1.7$ m s^{-1}. This may seem confusing, but the difficulty arises only when the motion involves a change of direction and we take an average value. If we are interested in describing the motion of a particle at a particular moment in time, the speed at that moment is the same as the magnitude of the velocity at that moment.

We now need to define average velocity more precisely, in terms of a mathematical equation, instead of our previous word equation. Suppose that at time t_1 a particle is at a point x_1 on the x-axis (Figure 2.1). At a later time t_2, the particle has moved to x_2. The displacement (the change in position) is $(x_2 - x_1)$, and the time taken is $(t_2 - t_1)$. The average velocity $\bar{v}$ is then

Figure 2.1

$$\bar{v} = \frac{x_2 - x_1}{t_2 - t_1}$$

The bar over v is the symbol meaning 'average'. As a shorthand, we can write $(x_2 - x_1)$ as Δx, where Δ (the Greek capital letter delta) means 'the change in'. Similarly, $t_2 - t_1$ is written as Δt. This gives us

$$\bar{v} = \frac{\Delta x}{\Delta t}$$

If x_2 were less than x_1, $(x_2 - x_1)$ and Δx would be negative. This would mean that the particle had moved to the left, instead of to the right as in Figure 2.1. The sign of the displacement gives the direction of particle motion. If Δx is negative, then the average velocity v is also negative. The sign of the velocity, as well as the sign of the displacement, indicates the direction of the particle's motion. This is because both displacement and velocity are vector quantities.

2.3 Describing motion by graphs

Position–time graphs

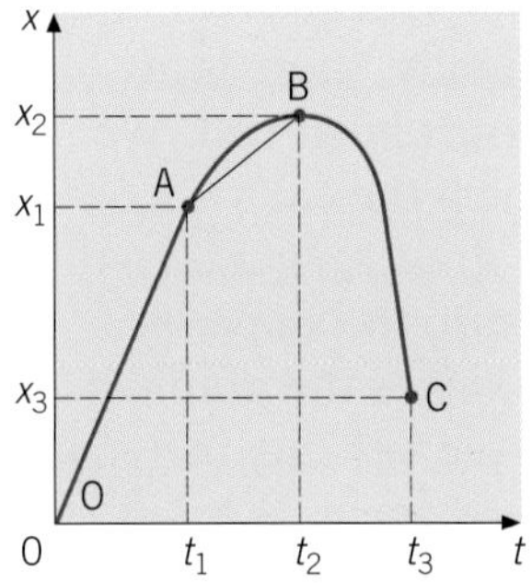

Figure 2.2 is a graph of position x against time t for a particle moving in a straight line. This curve gives a complete description of the motion of the particle. We can see from the graph that the particle starts at the origin O (at which $x = 0$) at time $t = 0$. From O to A the graph is a straight line: the particle is covering equal distances in equal periods of time. This represents a period of *uniform velocity*. The average velocity during this time is $(x_1 - 0)/(t_1 - 0)$. Clearly, this is the gradient of the straight-line part of the graph between O and A. Between A and B the particle is slowing down, because the distances travelled in equal periods of time are getting smaller. The average velocity during this period is $(x_2 - x_1)/(t_2 - t_1)$. On the graph, this is represented by the gradient of the straight line joining A and B. At B, for a moment, the particle is at rest, and after B it has reversed its direction and is

Figure 2.2

heading back towards the origin. Between B and C the average velocity is $(x_3 - x_2)/(t_3 - t_2)$. Because x_3 is less than x_2, this is a negative quantity, indicating the reversal of direction.

Calculating the average velocity of the particle over the relatively long intervals t_1, $(t_2 - t_1)$ and $(t_3 - t_2)$ will not, however, give us the complete description of the motion. To describe the motion exactly, we need to know the particle's velocity at every instant. We introduce the idea of **instantaneous velocity**. To define instantaneous velocity we make the intervals of time over which we measure the average velocity shorter and shorter. This has the effect of approximating the curved displacement–time graph by a series of short straight-line segments. The approximation becomes better the shorter the time interval, as illustrated in Figure 2.3. Eventually, in the case of extremely small time intervals (mathematically we would say 'infinitesimally small'), the straight-line segment has the same direction as the tangent to the curve. This limiting case gives the instantaneous velocity as the slope of the tangent to the displacement–time curve.

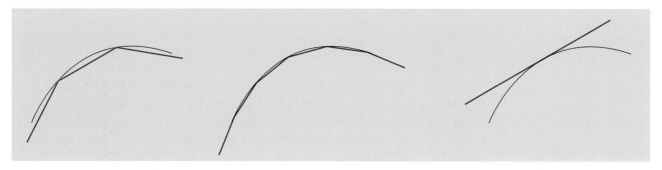

Figure 2.3

Displacement–time and velocity–time graphs

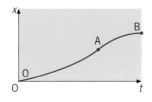

Figure 2.4

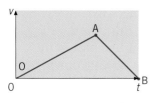

Figure 2.5

Figure 2.4 is a sketch graph showing how the displacement of a car, travelling along a straight test track, varies with time. We interpret this graph in a descriptive way by noting that between O and A the distances travelled in equal intervals of time are progressively increasing: that is, the velocity is increasing as the car is accelerating. Between A and B the distances for equal time intervals are decreasing; the car is slowing down. Finally, there is no change in position, even though time passes, so the car must be at rest. We can use Figure 2.4 to deduce the details of the way in which the car's instantaneous velocity v varies with time. To do this, we draw tangents to the curve in Figure 2.4 at regular intervals of time, and measure the slope of each tangent to obtain values of v. The plot of v against t gives the graph in Figure 2.5. This confirms our descriptive interpretation: the velocity increases from zero to a maximum value, and then decreases to zero again. We will look at this example in more detail on pages 31–32, where we shall see that the area under the velocity–time graph in Figure 2.5 gives the displacement x.

2.4 Acceleration

We have used the word *accelerating* in describing the increase in velocity of the car in the previous section. **Acceleration** is a measure of the rate at which the velocity of the particle is changing. **Average acceleration** is defined by the word equation

$$average\ acceleration = \frac{change\ in\ velocity}{time\ taken}$$

The unit of acceleration is the unit of velocity (the metre per second) divided by the unit of time (the second), giving the metre per second per second ($m\ s^{-2}$). In symbols, this equation is

$$\bar{a} = \frac{v_2 - v_1}{t_2 - t_1} = \frac{\Delta v}{\Delta t}$$

where v_1 and v_2 are the velocities at time t_1 and t_2 respectively. To obtain the **instantaneous acceleration**, we take extremely small time intervals, just as we did when defining instantaneous velocity. Because it involves a change in velocity (a vector quantity), acceleration is also a vector quantity: we need to specify both its magnitude and its direction.

We can deduce the acceleration of a particle from its velocity–time graph by drawing a tangent to the curve and finding the slope of the tangent. Figure 2.6 shows the result of doing this for the car's motion described by Figure 2.4 (the displacement–time graph) and Figure 2.5 (the velocity–time graph). The car accelerates at a constant rate between O and A, and then decelerates (that is, slows down) uniformly between A and B.

An acceleration with a very familiar value is the acceleration of free fall near the Earth's surface (see page 27): this is $9.81\ m\ s^{-2}$, often approximated to $10\ m\ s^{-2}$. To illustrate the range of values you may come across, some accelerations are summarised in Table 2.2.

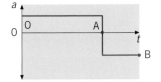

Figure 2.6

Table 2.2 Examples of accelerations

	acceleration/$m\ s^{-2}$
due to circular motion of electron around nucleus	9×10^{26}
car crash	1×10^3
free fall on Earth	10
family car	2
free fall on Moon	2
at equator, due to rotation of Earth	3×10^{-2}
due to circular motion of Earth around Sun	6×10^{-5}

Examples

1 A sports car accelerates along a straight test track from rest to
70 km h^{-1} in 6.3 s. What is its average acceleration?

First convert the data into consistent units. 70 km (7.0×10^4 m)
in 1 hour (3.6×10^3 s) is 19 m s^{-1}. Since average acceleration
is (change of velocity)/(time taken), the acceleration is
19/6.3 = **3.0 m s^{-2}**.

2 A railway train, travelling along a straight track, takes 1.5 minutes
to come to rest from a speed of 115 km h^{-1}. What is its average
acceleration while braking?

115 km h^{-1} is 31.9 m s^{-1}, and 1.5 minutes is 90 s. The average
acceleration is (change of velocity)/(time taken)
= −31.9/90 = **−0.35 m s^{-2}**.

Note that the acceleration is a negative quantity because the change
of velocity is negative: the final velocity is less than the initial. A
negative acceleration is often called a deceleration.

Now it's your turn

1 A sprinter, starting from the blocks, reaches his full speed of
9.0 m s^{-1} in 1.5 s. What is his average acceleration?

2 A car is travelling at a speed of 25 m s^{-1}. At this speed, it is capable of
accelerating at 1.8 m s^{-2}. How long would it take to accelerate from
25 m s^{-1} to the speed limit of 31 m s^{-1}?

Sections 2.1–2.4 Summary

- Speed is a scalar quantity, and is described by magnitude only.
  Velocity is a vector, and requires magnitude and direction. Speed is
  the magnitude of velocity. Displacement is the change in position of a
  particle, and is a vector quantity.
- Average speed is defined by: (*distance covered*)/(*time taken*)
- Average velocity is defined by: (*displacement*)/(*time taken*) or $\Delta x/\Delta t$
- The instantaneous velocity is the average velocity measured over an
  infinitesimally short time interval.
- Average acceleration is defined by: (*change in velocity*)/(*time taken*)
  or $\Delta v/\Delta t$
- Acceleration is a vector. Instantaneous acceleration is the average
  acceleration measured over an infinitesimally short time interval.
- Velocity may be deduced from the gradient of a displacement–time
  graph.
- Acceleration may be deduced from the gradient of a velocity–time
  graph.

1 At an average speed of 24 km h^{-1}, how many kilometres will a cyclist travel in 75 minutes?

2 An aircraft travels 1600 km in 2.5 hours. What is its average speed, in m s^{-1}?

3 Does a car speedometer register speed or velocity? Explain.

4 An aircraft travels 1400 km at a speed of 700 km h^{-1}, and then runs into a headwind that reduces its speed over the ground to 500 km h^{-1} for the next 800 km. What is the total time for the flight? What is the average speed of the aircraft?

5 A sports car can stop in 6.1 s from a speed of 110 km h^{-1}. What is its acceleration?

6 Can the velocity of a particle change if its speed is constant? Can the speed of a particle change if its velocity is constant? If the answer to either question is 'yes', give examples.

2.5 Uniformly accelerated motion

At the end of Sections 2.5–2.7 you should be able to:
- derive, from the definitions of velocity and acceleration, equations which represent uniformly accelerated motion in a straight line
- note the connections between the equations of uniformly accelerated motion and the graphs of displacement and velocity against time, including the use of the area under a velocity–time graph to find the displacement
- solve problems using equations which represent uniformly accelerated motion in a straight line, including the motion of bodies falling in a uniform gravitational field without air resistance
- explain how experiments carried out by Galileo overturned Aristotle's ideas of motion
- recall that the weight of a body is equal to the product of its mass and the acceleration of free fall g
- describe an experiment to determine the acceleration of free fall g using a falling body
- describe qualitatively the motion of bodies falling in a uniform gravitational field with air resistance (drag)
- use and explain term *terminal velocity*
- describe and explain motion due to a uniform velocity in one direction and a uniform acceleration in a perpendicular direction.

Having defined displacement, velocity and acceleration, we shall use the definitions to derive a series of equations, called the *kinematic equations*, which can be used to give a complete description of the motion of a particle in a straight line. The mathematics will be simplified if we deal with situations in which the acceleration does not vary with time; that is, the acceleration is uniform (or constant). This approximation applies for many practical cases. However, there are two important types of motion for which the kinematic equations do not apply: circular motion, and the oscillatory motion called simple harmonic motion.

Think about a particle moving along a straight line with constant acceleration a. Suppose that its initial velocity, at time $t = 0$, is u. After a further time t its velocity has increased to v. From the definition of acceleration as (change in velocity)/(time taken), we have $a = (v - u)/t$ or, re-arranging,

$$v = u + at$$

From the definition of average velocity $\bar{v}$ as (distance travelled)/(time taken), over the time t the distance travelled s will be given by the average velocity multiplied by the time taken, or

$$s = \bar{v}t$$

The average velocity $\bar{v}$ is written in terms of the initial velocity u and final velocity v as

$$\bar{v} = \frac{u + v}{2}$$

and, using the previous equation for v,

$$\bar{v} = (u + u + at)/2 = u + at/2$$

Substituting this we have

$$s = ut + \frac{1}{2}at^2$$

The right-hand side of this equation is the sum of two terms. The ut term is the distance the particle would have travelled in time t if it had been travelling with a constant speed u, and the $\frac{1}{2}at^2$ term is the additional distance travelled as a result of the acceleration.

The equation relating the final velocity v, the initial velocity u, the acceleration a and the distance travelled s is

$$v^2 = u^2 + 2as$$

If you wish to see how this is obtained from previous equations, see the Maths Note below.

Maths Note

From $v = u + at$,

$$t = (v - u)/a$$

Substitute this in $s = ut + \frac{1}{2}at^2$,

$$s = u(v - u)/a + \frac{1}{2}a(v - u)^2/a^2$$

Multiplying both sides by $2a$ and expanding the terms,

$$2as = 2uv - 2u^2 + v^2 - 2uv + u^2$$

or $v^2 = u^2 + 2as$

The four equations relating the various quantities which define the motion of the particle in a straight line in uniformly accelerated motion are

$$v = u + at$$
$$s = ut + \frac{1}{2}at^2$$
$$v^2 = u^2 + 2as$$
$$\bar{v} = (u + v)/2$$

In these equations u is the initial velocity, v is the final velocity, $\bar{v}$ is the average velocity, a is the acceleration, s is the distance travelled, and t is the time taken.

In solving problems involving kinematics, it is important to understand the situation before you try to substitute numerical values into an equation. Identify the quantity you want to know, and then make a list of the quantities you know already. This should make it obvious which equation is to be used.

Free fall acceleration

A very common example of uniformly accelerated motion is when a body falls freely near the Earth's surface. Because of the gravitational attraction of the Earth, all objects fall with the same uniform acceleration. This acceleration is called the **acceleration of free fall**, and is represented by the symbol g. It has a value of 9.81 m s^{-2}, and is directed downwards. For completeness, we ought to qualify this statement by saying that the fall must be in the absence of air resistance, but in most situations this can be assumed to be true.

One reason why the acceleration of free fall is important is that it provides the link between the mass m and the weight W of a body. The relationship is

$$W = mg$$

We will meet this equation again in Chapter 4. At present you need only remember the relationship.

The acceleration of free fall may be determined by an experiment in which the time of fall t of a body between two points a distance s apart is measured. If the body falls from rest, we can use the second of the equations for uniformly accelerated motion in the form

$$g = 2s/t^2$$

to calculate the value of g. Note that, because the time of fall is likely to be only a few tenths of a second, precise timing to one-hundredth of a second is required. This is achieved by the switching of light gates by the falling object (Figure 2.8). The light gates are connected to an electronic timer.

Until the sixteenth century, the idea of the acceleration of a falling body was not fully appreciated. The Greek philosopher Aristotle (384–322 BC) thought that heavier bodies fell faster than light ones. This idea was a consequence of the effect of air resistance on light objects with a large surface area, such as feathers. However, Galileo Galilei (1564–1642) suggested that, in the absence of resistance, all bodies would fall with the same constant acceleration. He showed mathematically that, for a body falling from rest, the distance travelled is

Figure 2.7 Strobo-flash photograph of objects in free fall.

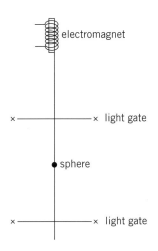

Figure 2.8 Determination of the acceleration of free fall

Figure 2.9 Galileo in his study

Figure 2.10 Leaning Tower of Pisa

Figure 2.11 A parachutist about to land

proportional to the square of the time. Galileo tested the relation experimentally by timing the fall of objects from various levels of the Leaning Tower of Pisa (Figure 2.10). This is the relation we have derived as $s = ut + \frac{1}{2}at^2$. For a body starting from rest, $u = 0$ and $s = \frac{1}{2}at^2$. That is, the distance is proportional to time squared.

We have mentioned that, in most situations, air resistance can be neglected. In fact, there are some applications in which this resistance is most important. One is the case of the fall of a parachutist, where air resistance plays a vital part. The velocity of a body falling through a resistive fluid (a liquid or a gas) does not increase indefinitely, but eventually reaches a maximum velocity, called the **terminal velocity**. The force due to air resistance (drag) increases with speed. When this resistive force has reached a value equal and opposite to the weight of the falling body, the body no longer accelerates and continues at uniform (constant) velocity. This is a case of motion with non-uniform acceleration. The acceleration starts off with a value of g, but decreases to zero at the time when the terminal velocity is achieved. Thus, raindrops and parachutists are normally travelling at a constant speed by the time they approach the ground (Figure 2.11).

Examples

1 A car increases its speed from $25\,\text{m s}^{-1}$ to $31\,\text{m s}^{-1}$ with a uniform acceleration of $1.8\,\text{m s}^{-2}$. How far does it travel while accelerating?

In this problem we want to know the distance s. We know the initial speed $u = 25\,\text{m s}^{-1}$, the final speed $v = 31\,\text{m s}^{-1}$, and the acceleration $a = 1.8\,\text{m s}^{-2}$. We need an equation linking s with u, v and a; this is

$$v^2 = u^2 + 2as$$

Substituting the values, we have $31^2 = 25^2 + 2 \times 1.8s$. Re-arranging, $s = (31^2 - 25^2)/(2 \times 1.8) = \textbf{93 m}$.

2 The average acceleration of a sprinter from the time of leaving the blocks to reaching his maximum speed of $9.0\,\mathrm{m\,s^{-1}}$ is $6.0\,\mathrm{m\,s^{-2}}$. For how long does he accelerate? What distance does he cover in this time?

In the first part of this problem, we want to know the time t. We know the initial speed $u = 0$, the final speed $v = 9.0\,\mathrm{m\,s^{-1}}$, and the acceleration $a = 6.0\,\mathrm{m\,s^{-2}}$. We need an equation linking t with u, v and a; this is

$$v = u + at$$

Substituting the values, we have $9.0 = 0 + 6.0t$. Re-arranging, $t = 9.0/6.0 = \mathbf{1.5\,s}$.

For the second part of the problem, we want to know the distance s. We know the initial speed $u = 0$, the final speed $v = 9.0\,\mathrm{m\,s^{-1}}$, and the acceleration $a = 6.0\,\mathrm{m\,s^{-2}}$; we have also just found the time $t = 1.5\,\mathrm{s}$. There is a choice of equations linking s with u, v, a and t. We can use

$$s = ut + \tfrac{1}{2}at^2$$

Substituting the values, $s = 0 + \tfrac{1}{2} \times 6.0 \times (1.5)^2 = \mathbf{6.8\,m}$.

Another relevant equation is $\bar{v} = \Delta x/\Delta t$. Here the average velocity $\bar{v}$ is given by $\bar{v} = (u + v)/2 = 9.0/2 = 4.5\,\mathrm{m\,s^{-1}}$. $\Delta x/\Delta t$ is the same as s/t, so $4.5 = s/1.5$, and $s = 4.5 \times 1.5 = \mathbf{6.8\,m}$ as before.

3 A cricketer throws a ball vertically upward into the air with an initial velocity of $18.0\,\mathrm{m\,s^{-1}}$. How high does the ball go? How long is it before it returns to the cricketer's hands?

In the first part of the problem, we want to know the distance s. We know the initial velocity $u = 18.0\,\mathrm{m\,s^{-1}}$ upwards and the acceleration $a = g = 9.81\,\mathrm{m\,s^{-2}}$ downwards. At the highest point the ball is momentarily at rest, so the final velocity $v = 0$. The equation linking s with u, v and a is

$$v^2 = u^2 + 2as$$

Substituting the values, $0 = (18.0)^2 + 2(-9.81)s$. Thus $s = -(18.0)^2/2(-9.81) = \mathbf{16.5\,m}$. Note that here the ball has an upward velocity but a downward acceleration, and that at the highest point the velocity is zero but the acceleration is not zero.

In the second part we want to know the time t for the ball's up-and-down flight. We know u and a, and also the overall displacement $s = 0$, as the ball returns to the same point at which it was thrown. The equation to use is

$$s = ut + \tfrac{1}{2}at^2$$

Substituting the values, $0 = 18.0t + \tfrac{1}{2}(-9.81)t^2$. Doing some algebra, $t(36.0 - 9.81t) = 0$. There are two solutions, $t = 0$ and $t = 36.0/9.81 = 3.7\,\mathrm{s}$. The $t = 0$ value corresponds to the time when the displacement was zero when the ball was on the point of leaving the cricketer's hands. The answer required here is $\mathbf{3.7\,s}$.

2.6 Graphs of the kinematic equations

It is often useful to represent the motion of a particle graphically, instead of by means of a series of equations. In this section we bring together the graphs which correspond to the equations we have already derived. We shall see that there are some important links between the graphs.

First, think about a particle moving in a straight line with constant velocity. Constant velocity means that the particle covers equal distances in equal intervals of time. A graph of displacement x against time t is thus a straight line, as in Figure 2.12. Here the particle has started at $x = 0$ and at time $t = 0$. The slope of the graph is equal to the magnitude of the velocity, since, from the definition of average velocity, $\bar{v} = (x_2 - x_1)/(t_2 - t_1) = \Delta x/\Delta t$. Because this graph is a straight line, the average velocity and the instantaneous velocity are the same. The equation describing the graph is $x = vt$.

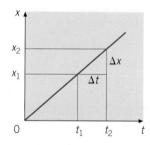

Figure 2.12

Now think about a particle moving in a straight line with constant acceleration. The particle's velocity will change by equal amounts in equal intervals of time. A graph of the magnitude v of the velocity against time t will be a straight line, as in Figure 2.13. Here the particle has started with velocity u at time $t = 0$. The slope of the graph is equal to the magnitude of the acceleration. The graph is a straight line showing that the acceleration is a constant. The equation describing the graph is $v = u + at$.

An important feature of the velocity–time graph is that we can deduce the displacement of the particle by calculating the area between the graph and the t-axis, between appropriate limits of time. Suppose we want to obtain the displacement of the particle between times t_1 and t_2 in Figure 2.13. Between these times the average velocity $\bar{v}$ is represented by the horizontal line AB. The area between the graph and the t-axis is equal to the area of the rectangle whose top edge is AB, or average velocity $\bar{v}$. This area is $\bar{v}\Delta t$. But, by the definition of average velocity ($\bar{v} = \Delta x/\Delta t$), $\bar{v}\Delta t$ is equal to the displacement Δx during the time interval Δt.

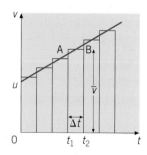

Figure 2.13

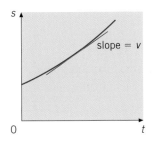

Figure 2.14

We can deduce the graph of displacement s against time t from the velocity–time graph by calculating the area between the graph and the t-axis for a succession of values of t. As shown in Figure 2.13, we can split the area up into a number of rectangles. The displacement at a certain time is then just the sum of the areas of the rectangles up to that time. Figure 2.14 shows the result of plotting the displacement s determined in this way against time t. It is a curve with a slope which increases the higher the value of t, indicating that the particle is accelerating. The slope at a particular time gives the magnitude of the instantaneous velocity. The equation describing Figure 2.14 is $s = ut + \frac{1}{2}at^2$.

Example

The displacement–time graph for a car on a straight test track is shown in Figure 2.15. Use this graph to draw velocity–time and acceleration–time graphs for the test run.

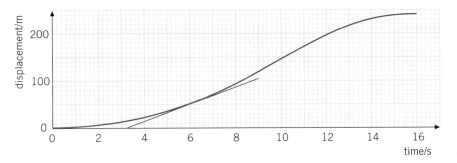

Figure 2.15 Displacement–time graph

We have already met this graph when we discussed the concepts of velocity and acceleration (Figure 2.4, page 22). In Figure 2.15 it has been re-drawn to scale, and figures have been put on the displacement and time axes. We find the magnitude of the velocity by measuring the gradient of the displacement–time graph. As an example, a tangent to the graph has been drawn at $t = 6.0$ s. The slope of this tangent is $18\,\text{m s}^{-1}$. If the process is repeated at different times, the following velocities are determined.

t/s	2	4	6	8	10	12	14	16
v/m s^{-1}	6	12	18	24	30	20	10	0

These values are plotted on the velocity–time graph of Figure 2.16. Check some of the values by drawing tangents yourself. *Hint*: When drawing tangents, use a mirror or a transparent ruler.

Figure 2.16 shows two straight-line portions. Initially, from $t = 0$ to $t = 10$ s, the car is accelerating uniformly, and from $t = 10$ s to $t = 16$ s it is decelerating. The acceleration is given by $a = \Delta v / \Delta t = 30/10 = 3\,\text{m s}^{-2}$ up to $t = 10$ s. Beyond $t = 10$ s the acceleration is $-30/6 = -5\,\text{m s}^{-2}$. (The minus sign shows that the car is decelerating.)

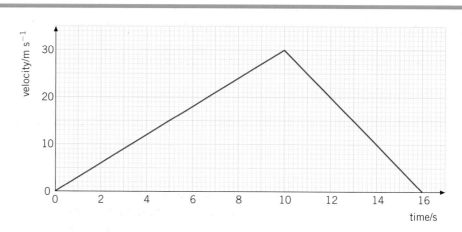

Figure 2.16 Velocity–time graph

The acceleration–time graph is plotted in Figure 2.17.

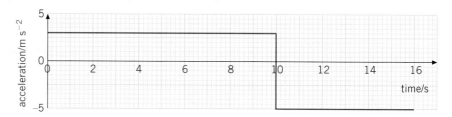

Figure 2.17 Acceleration–time graph

Finally, we can confirm that the area under a velocity–time graph gives the displacement. The area under the line in Figure 2.16 is

$$\left(\tfrac{1}{2} \times 10 \times 30\right) + \left(\tfrac{1}{2} \times 6 \times 30\right) = 240\,\text{m}$$

the value of s at $t = 16\,\text{s}$ on Figure 2.15.

Now it's your turn

In a test of a sports car on a straight track, the following readings of velocity v were obtained at the times t stated.

t/s	0	5	10	15	20	25	30	35
$v/\text{m s}^{-1}$	0	15	23	28	32	35	37	38

(a) On graph paper, draw a velocity–time graph and use it to determine the acceleration of the car at time $t = 5\,\text{s}$.
(b) Find also the total distance travelled between $t = 0$ and $t = 30\,\text{s}$.

Note: These figures refer to a case of non-uniform acceleration, which is more realistic than the previous example. However, the same rules apply: the acceleration is given by the slope of the velocity–time graph at the relevant time, and the distance travelled can be found from the area under the graph.

2.7 Two-dimensional motion under a constant force

Figure 2.18 Cricketer bowling the ball.

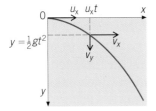

Figure 2.19

So far we have been dealing with motion along a straight line; that is, one-dimensional motion. We now think about the motion of particles moving in paths in two dimensions. We shall need to make use of ideas we have already learnt regarding vectors in Chapter 1. The particular example we shall take is where a particle moves in a plane under the action of a constant force. An example is the motion of a ball thrown at an angle to the vertical (Figure 2.18), or an electron moving at an angle to an electric field. In the case of the ball, the constant force acting on it is its weight. For the electron, the constant force is the force provided by the electric field.

This topic is often called **projectile motion**. Galileo first gave an accurate analysis of this motion. He did so by splitting the motion up into its vertical and horizontal components, and considering these separately. The key is that the two components can be considered independently.

As an example, think about a particle sent off in a horizontal direction and subject to a vertical gravitational force (its weight). As before, air resistance will be neglected. We will analyse the motion in terms of the horizontal and vertical components of velocity. The particle is projected at time $t = 0$ at the origin of a system of x, y co-ordinates (Figure 2.19) with velocity u_x in the x-direction. Think first about the particle's vertical motion (in the y-direction). Throughout the motion, it has an acceleration of g (the acceleration of free fall) in the y-direction. The initial value of the vertical component of velocity is $u_y = 0$. The vertical component increases continuously under the uniform acceleration g. Using $v = u + at$, its value v_y at time t is given by $v_y = gt$. Also at time t, the vertical displacement y downwards is given by $y = \frac{1}{2}gt^2$. Now for the horizontal motion (in the x-direction): here the acceleration is zero, so the horizontal component of velocity remains constant at u_x. At time t the horizontal displacement x is given by $x = u_x t$. To find the velocity of the particle at any time t, the two components v_x and v_y must be added vectorially. The direction of the resultant vector is the direction of motion of the particle. The curve traced out by a particle subject to a constant force in one direction is a **parabola**.

Figure 2.20 Water jets from a garden sprinkler showing a parabola-shaped spray.

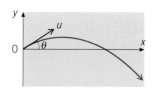

Figure 2.21

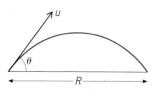

Figure 2.22

If the particle had been sent off with velocity u at an angle θ to the horizontal, as in Figure 2.21, the only difference to the analysis of the motion is that the initial y-component of velocity is $u \sin \theta$. In the example illustrated in Figure 2.21 this is upwards. Because of the downwards acceleration g, the y-component of velocity decreases to zero, at which time the particle is at the crest of its path, and then increases in magnitude again but this time in the opposite direction. The path is again a parabola.

For the particular case of a particle projected with velocity u at an angle θ to the horizontal from a point on level ground (Figure 2.22), the range R is defined as the distance from the point of projection to the point at which the particle reaches the ground again. We can show that R is given by

$$R = \frac{(u^2 \sin 2\theta)}{g}$$

For details, see the Maths Note below.

Maths Note

Suppose that the particle is projected from the origin ($x = 0$, $y = 0$). We can interpret the range R as being the horizontal distance x travelled at the time t when the value of y is again zero. The equation which links displacement, initial speed, acceleration and time is $s = ut + \frac{1}{2}at^2$. Adapting this for the vertical component of the motion, we have

$$0 = (u \sin \theta)t - \tfrac{1}{2}gt^2$$

The two solutions of this equation are $t = 0$ and $t = (2u \sin \theta)/g$. The $t = 0$ case is when the particle was projected; the second is when it returns to the ground at $y = 0$. We use this second value of t with the horizontal component of velocity $u \cos \theta$ to find the distance x travelled (the range R). This is

$$x = R = (u \cos \theta)t = (2u^2 \sin \theta \cos \theta)/g$$

There is a trigonometric relation $\sin 2\theta = 2 \sin \theta \cos \theta$, use of which puts the range expression in the required form

$$R = (u^2 \sin 2\theta)/g$$

We can see that R will have its maximum value for a given speed of projection u when $\sin 2\theta = 1$, or $2\theta = 90°$, or $\theta = 45°$. The value of this maximum range is $R_{max} = u^2/g$.

Examples

1 A stone is thrown from the top of a vertical cliff, 45 m high above level ground, with an initial velocity of 15 m s^{-1} in a horizontal direction (Figure 2.23). How long does it take to reach the ground? How far from the base of the cliff is it when it reaches the ground?

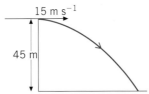

Figure 2.23

To find the time t for which the stone is in the air, work with the vertical component of the motion, for which we know that the initial component of velocity is zero, the displacement $y = 45$ m, and the acceleration a is 9.81 m s^{-2}. The equation linking these is $y = \frac{1}{2}gt^2$. Substituting the values, we have $45 = \frac{1}{2} \times 9.81t^2$. This gives $t = \sqrt{(2 \times 45/9.81)} = \mathbf{3.0\,s}$.

For the second part of the question, we need to find the horizontal distance x travelled in the time t. Because the horizontal component of the motion is not accelerating, x is given simply by $x = u_x t$. Substituting the values, we have $x = 15 \times 3.0 = \mathbf{45\,m}$.

2 An electron, travelling with a velocity of 2.0×10^7 m s^{-1} in a horizontal direction, enters a uniform electric field. This field gives the electron a constant acceleration of 5.0×10^{15} m s^{-2} in a direction perpendicular to its original velocity (Figure 2.24). The field extends for a horizontal distance of 60 mm. What is the magnitude and direction of the velocity of the electron when it leaves the field?

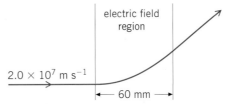

Figure 2.24

The horizontal motion of the electron is not accelerated. The time t spent by the electron in the field is given by $t = x/u_x = 60 \times 10^{-3}/2.0 \times 10^7 = 3.0 \times 10^{-9}$ s. When the electron enters the field, its vertical component of velocity is zero; in time t, it has been accelerated to $v_y = at = 5.0 \times 10^{15} \times 3.0 \times 10^{-9} = 1.5 \times 10^7$ m s^{-1}. When the electron leaves the field, it has a horizontal component of velocity $v_x = 2.0 \times 10^7$ m s^{-1}, unchanged from the initial value u_x. The vertical component is $v_y = 1.5 \times 10^7$ m s^{-1}. The resultant velocity v is given by $v = \sqrt{(v_x^2 + v_y^2)} = \sqrt{[(2.0 \times 10^7)^2 + (1.5 \times 10^7)^2]} = \mathbf{2.5 \times 10^7\,m\,s^{-1}}$. The direction of this resultant velocity makes an angle θ to the horizontal, where θ is given by $\tan \theta = v_y/v_x = 1.5 \times 10^7/2.0 \times 10^7$. The angle θ is **36.9°**.

Now it's your turn

1 A ball is thrown horizontally from the top of a tower 30 m high and lands 15 m from its base (Figure 2.25). What is the ball's initial speed?

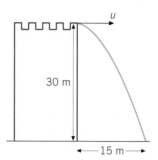

Figure 2.25

2 A football is kicked on level ground at a velocity of 15 m s^{-1} at an angle of 30° to the horizontal (Figure 2.26). How far away is the first bounce?

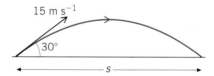

Figure 2.26

Sections 2.5–2.7 Summary

▪ The equations for a body moving in a straight line with uniform acceleration are:

$$v = u + at \qquad\qquad v^2 = u^2 + 2as$$
$$s = ut + \tfrac{1}{2}at^2 \qquad\qquad \bar{v} = (u + v)/2$$

▪ Objects falling freely near the surface of the Earth in the absence of air resistance, experience the same acceleration, the acceleration of free fall g, which has the value $g = 9.81$ m s^{-2}.

▪ The acceleration of free fall provides the link between the mass m and the weight W of a body: $W = mg$

▪ The gradient of a displacement–time graph gives the velocity of a particle, and the gradient of a velocity–time graph gives its acceleration. The area between a velocity–time graph and the time axis gives the displacement.

▪ The motion of projectiles is analysed in terms of two independent motions at right angles. The horizontal component of the motion is at a constant velocity, while the vertical motion is subject to a constant acceleration g.

Sections 2.5–2.7 Questions

1 A car accelerates from $5.0 \, \text{m s}^{-1}$ to $20 \, \text{m s}^{-1}$ in 6.0 s. Assuming uniform acceleration, how far does it travel in this time?

2 If a raindrop were to fall from a height of 1 km, with what velocity would it hit the ground if there were no air resistance?

3 Traffic police can estimate the speed of vehicles involved in accidents by the length of the marks made by skidding tyres on the road surface. It is known that the maximum deceleration that a car can attain when braking on a normal road surface is about $9 \, \text{m s}^{-2}$. In one accident, the tyre-marks were found to be 125 m long. Estimate the speed of the vehicle before braking.

4 On a theme park ride, a cage is travelling upwards at constant speed. As it passes a platform alongside, a passenger drops coin A through the cage floor. At exactly the same time, a person standing on the platform drops coin B from the platform.

Which coin, A or B (if either), reaches the ground first? Which (if either) has the greater speed on impact?

5 William Tell was faced with the agonising task of shooting an apple from his son Jemmy's head. Assume that William is placed 25 m from Jemmy; his crossbow fires a bolt with an initial speed of $45 \, \text{m s}^{-1}$. The crossbow and apple are on the same horizontal line. At what angle to the horizontal should William aim so that the bolt hits the apple?

6 The position of a sports car on a straight test track is monitored by taking a series of photographs at fixed time intervals. The following record of position x was obtained at the stated times t.

t/s	0	0.5	1.0	1.5	2.0	2.5	3.0	3.5	4.0	4.5	5.0
x/m	0	0.4	1.8	4.2	7.7	12.4	18.3	25.5	33.9	43.5	54.3

On graph paper, draw a graph of x against t. Use your graph to obtain values for the velocity v of the car at a number of values of t. Draw a second graph of v against t. From this graph, what can you deduce about the acceleration of the car?

2.8 Car safety

At the end of Section 2.8, you should be able to:
- define *thinking distance*, *braking distance* and *stopping distance*, and analyse and solve problems that use these terms
- describe the factors that affect thinking distance and braking distance
- describe and explain how air bags, seat belts and crumple zones reduce impact forces in accidents
- describe how air bags work
- describe how the trilateration technique is used in the Global Positioning System (GPS) for cars.

A study of car safety gives us an opportunity to see how a lot of the physics we have learned up to now is applied to real life. We shall think first of **active safety**, in which the risk of damage to the car and injury to the occupants is at least partially under the control of the driver.

Stopping a car

The process of stopping a moving car can be split into two parts, thinking (or reacting) and braking. The times taken for each part are the thinking time (often called the reaction time) and the braking time respectively. They add up to the total time taken to stop the car, the stopping time.

The thinking time is the time taken while the driver, having seen a hazard ahead, decides whether to apply the brakes to bring the car to a halt. The braking time is the time taken between the instant of the foot being applied to the brake pedal and the car stopping. The thinking time for an alert driver is less than a second; 0.7 s is often taken as the standard reaction time. This time is affected by whether the driver is tired, so that the brain thinks more slowly and the reaction time is increased. If the driver is under the influence of drink or drugs the effect will certainly be to decrease awareness and judgement of hazards. In-car distractions, in particular the use of mobile telephones, will also reduce the driver's attention to hazards. Braking times are affected by different factors, such as the condition of the brakes and tyres, the type of road surface and the mass of the car and load. For example, worn brake pads do not grip so well on the discs as new ones, so that the brakes must be applied for a longer time. Some tyres are designed for greater fuel economy by reducing the rolling friction; this means that they will have a less satisfactory grip on wet roads. Similarly, different types of road surface have different degrees of grip. If a car is heavily loaded, its momentum (which is proportional to mass) will be greater at a given speed than a lightly-loaded car moving at the same speed. This means that the braking time will be greater, since the change in momentum from the speed at which the brakes were applied to rest is equal to the product of the average braking force and the braking time.

In relation to an accident occurring, the stopping distance, the sum of the distances travelled by the car during the thinking and braking times (the thinking and braking distances) is more important than the corresponding stopping time.

The definitions are

Thinking (reaction) distance is the distance travelled between the recognition of a hazard by the driver and the first application of the brakes.

Braking distance is the distance travelled between the first application of the brakes and the car coming to a halt.

Stopping distance is the total distance travelled between the first recognition of a hazard by the driver and the car coming to a halt. It is equal to the sum of the thinking distance and the braking distance.

In analysing problems about thinking distance, braking distance and stopping distance we make use of simple kinematics equations from Section 2.5.

Because the car is travelling at uniform speed during the thinking time, we can use the equation

$$s = ut + \tfrac{1}{2}at^2$$

with the acceleration a equal to zero, as the brakes have not been applied yet; the driver is still thinking about doing so. This means that the thinking distance s_r is given by

$$s_r = ut_r$$

where u is the steady initial speed of the car and t_r is the thinking (reaction) time. Since the thinking time is a constant, the thinking distance is directly proportional to the speed of the car.

Assume that the brakes apply a constant retarding force once they have been applied. This means that the deceleration is constant. Starting from the equation

$$v^2 = u^2 + 2as$$

with a deceleration (negative acceleration) a, an initial speed of u and a final speed of zero (when the car has come to rest), the braking distance s_b is given by

$$s_b = u^2/2a$$

Here we see that, because the braking distance is proportional to the square of the speed of the car when the brakes are first applied, s_b increases rapidly with u.

Example

The UK Highway Code gives the following figures for typical thinking distances s_r and braking distances s_b for a car travelling at various speeds u.

u/m.p.h	20	30	40	50	60	70
u/km hr^{-1}	32	48	64	80	96	112
s_r/m	6	9	12	15	18	21
s_b/m	6	14	24	38	55	75

Calculate the value of the stopping distance s for each speed u. On the same axes, plot graphs of s_r, s_b and s against u. Comment on the forms of the graphs.

The values of s are obtained from $s = s_r + s_b$. They are

s/m	12	23	36	53	73	96

The graphs are shown in Figure 2.27.

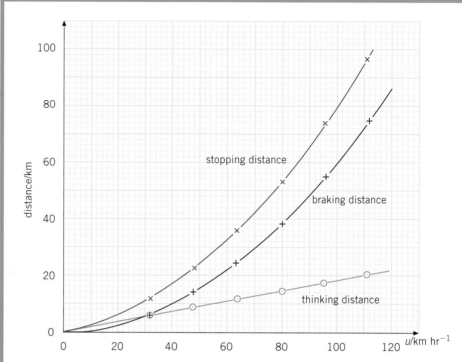

Figure 2.27 Graphs of s_r, s_b and s against u

The graph of s_r against u is a straight line through the origin, showing that the thinking distance is directly proportional to u. This is as predicted by the equation

$$s_r = ut_r$$

in which the thinking time t_r is a constant. Thus the thinking time is the gradient of the graph.

If we were to calculate the values of u^2, and plot the values of s_b against u^2, we would obtain a straight line through the origin. (Try it!) This straight-line graph shows that the braking distance s_b is directly proportional to u^2, as predicted by the equation

$$s_b = u^2/2a$$

At low speeds, the thinking distance is actually greater than the braking distance. But for speeds greater than 32 km hr^{-1} (20 m.p.h.) the braking distance curve increases quickly and rises well above the thinking distance line. Because the stopping distance is defined as the sum of the thinking and braking distances, its curve must lie above the graphs of either thinking or braking distance. Its graph emphasises how quickly the stopping distance increases with increasing speed, and shows the need to drive at a speed that is consistent with the hazards likely to be encountered. Thus, the speed limit in built-up areas in the UK is set at 30 m.p.h., at which the stopping distance is 23 metres, or about six car lengths. Drivers at motorway speeds of 70 m.p.h. should realise that their stopping distance is 96 metres, or about 24 car lengths.

▶▶

Now it's your turn

1   Use the data in the material from the Highway Code, or take measurements from relevant graphs, to calculate **(i)** the thinking time for the average driver, **(ii)** the braking deceleration of the average car.

2   The UK Highway Code recommends that a driver on motorways travelling at speeds of 70 m.p.h. should leave a distance of at least five car lengths between his/her car and the car in front. Does this make sense, given that the stopping distance at 70 m.p.h. is as much as 24 car lengths?

The reduction of impact forces

In this section we shall be dealing with **passive safety**, in which features of the design of a vehicle help to reduce the risk of injury to the occupants. Remember that, by Newton's second law, the greater the deceleration of a car in a collision, the greater the forces involved. Also, the greater the speed of a car, the greater is the kinetic energy of the car. In a head-on collision in which the speed is reduced to zero, this energy must be dissipated. It is important that as little as possible of this energy is imparted to the passengers.

Most cars on the European market are fitted with at least two **air bags**. An air bag is essentially a large gas-filled balloon which cushions a passenger from impact with parts of the car, usually the steering wheel or windscreen. The balloon inflates when the car decelerates very rapidly, as in the case of a head-on collision. An air bag is intended to complement the restraint provided by a seat belt. The air bag is not a sealed balloon, but has built-in vent holes, so that during a head-on collision in which the passenger is being thrown forward, gas is forced out of the vents, causing a gradual deflation. It is this which provides the cushioning effect. The volumes of the bag, and the size of the vents, are designed to dissipate the occupant's energy over a period of time, and also to spread the deceleration force over a much wider area than that provided by the seat belt.

The air bag is triggered by a signal from an accelerometer, a sensor that detects rapid deceleration such as might occur in a collision. This signal ignites a gas generator, which fills the air bag, a large plastic balloon, very rapidly. Clearly, the design of the accelerometer is critical. It must not set off the inflation process during the normal deceleration of the car during braking, nor during relatively small impacts that might involve no more than small dents to the bodywork. Modern sensors are small circuit chips which have tiny built-in mechanical elements. These elements move if the car decelerates at more than a certain rate. The change in geometry is detected, and an electrical signal fires the gas generator. Currently, the gases used in many bags are nitrogen or a mixture of nitrogen and argon. Air bags must inflate very rapidly if they are to be any use in stopping a passenger from hitting the interior of a car.

Figure 2.28 Three-point car seat belt

The two air bags referred to are commonly fitted in the centre of the steering wheel and in the dashboard, so that they help to protect the driver and the front-seat passenger in head-on or oblique collisions. Some collisions, however, occur from the side, and air bags may also be fitted in doors or in the outer edges of the front seatbacks.

By UK law, the driver and all passengers must wear a **seat belt**. (Certain classes of passenger, for example holders of medical exemption certificates, are excused.) The type fitted to front seats, and to the outer rear seats, is a three-point belt (Figure 2.28). Some centre rear seats are fitted with a lap belt that goes only over the lower waist.

Occupants using belts are restrained from hitting parts of the car, such as the steering wheel, windscreen or dashboard, if it is brought to a sudden stop in an accident. They also stop occupants being thrown out of the car if the doors burst open as a result of an impact. Most belts are fitted with a simple mechanical accelerometer that locks the belt in position if it is pulled out of the reel rapidly, as would occur if the car were suddenly stopped and the wearer of the belt moved quickly forward. Some have a pre-tensioner connected to the air bag accelerometer; this locks the belt in position. Clearly, wearing a seat belt will not guarantee that a passenger does not suffer injury; indeed, when they were first introduced, it was argued that injuries could be caused by the belt itself. However, the belt will very much reduce the risk of impact with hard parts of the car, and when used in combination with an air bag, the degree of protection is greatly increased.

Most cars incorporate **crumple zones** into their design. These are parts of the structure that are designed to compress during an impact so as to absorb the energy of the still-moving vehicle. These zones are generally included at the front and rear of the car. Figure 2.29 shows the result of a minor accident where the zone at the front of the car has crumpled, but the passenger compartment is completely free of damage.

Figure 2.29 Crumple zones in action

In addition to including these features, the part where the passengers sit (the passenger cage) is reinforced to prevent distortion of the cage and the intrusion of components from the crumple zones.

The Global Positioning System (GPS)

Many cars are fitted with a link to the Global Positioning System (GPS). The unit in the car is a computer that allows the driver to determine the current position of the car and to receive directions for the journey.

The system is based on a number of satellites which orbit about the centre of the Earth. The orbits are arranged so that at least six satellites are within the line of sight of almost any point on Earth. Actually, the system will work with only four satellites visible at the same time; if we are dealing with the location of points at the same, known level (for example, ships at sea) only three satellites are required. The GPS receiver in the car receives signals from four or more of the satellites; the computer in the receiver solves the signals for three spatial co-ordinates (x, y, z) and the local time t, The computer then translates the co-ordinates into position on the map display unit.

Each GPS satellite continually transmits microwave signals containing information on its own position, and on the time of the signal. Because the signals travel at a known speed (the speed c of electromagnetic radiation in a vacuum through space and a speed $c - \Delta c$ through the Earth's atmosphere) the in-car receiver can calculate, from the arrival time of the signal, the distance from the satellite. Using information from four or more satellites, the receiver's computer calculates the distances from each of these satellites, and then determines the exact position of the car using a trigonometrical technique called *trilateration*.

Trilateration should not be confused with triangulation, a technique widely used in map-making. In trilateration the measured *distances* between three reference points (the satellites) and the position of the receiver in the car are used to locate the car. In triangulation, positions are determined using two or more *angle* measurements and a known distance. We will demonstrate the trilateration method using a simple two-dimensional example. We want to find the location of a receiver at a point in the same plane as three satellites, from which it is receiving signals. Referring to Figure 2.30, the GPS receiver is at the point A.

The receiver picks up a signal from satellite S_1, the first of the three reference points. The receiver works out that the distance between it and this satellite is r_1. The computer's conclusion is that A must lie somewhere on the circumference of a circle of radius r_1 centred on S_1. This circle is shown in the upper diagram of Figure 2.30. At the same time the receiver's computer has been dealing with a signal from satellite S_2, the second reference point. The computer works out that the distance between it and S_2 is r_2, so that A must also lie somewhere on the circumference of a circle of radius r_2 centred on S_2. That is, it must lie at one of the two points A or B in which the two circles intersect. This is illustrated in the central diagram. To decide between these alternatives, the GPS computer makes use of information from a third reference point, the satellite S_3. The signal from this satellite tells the computer that it is at a distance r_3 away, so that it must lie on a circle of radius r_3 centred on S_3 as well as on each of the other two circles. The single point of intersection of the three circles gives the location of the GPS receiver at A, as shown in the lower diagram.

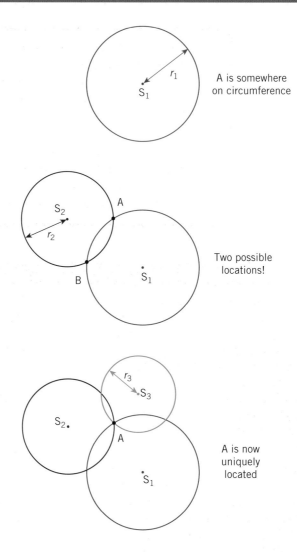

Figure 2.30

Section 2.8 Summary

▨ *Active safety* relates to safety procedures that are at least partly under the control of the driver of a vehicle; *passive safety* relates to features of the design of a vehicle intended to promote reduction of injury through impact forces.

▨ The total stopping distance of a vehicle is made up of the distance the vehicle travels between the identification of a hazard by the driver and the application of the brakes (the thinking distance) and the distance the vehicle travels between the first application of the brakes and coming to rest (the braking distance).

▨ The thinking distance is proportional to the speed of the vehicle; the braking distance is proportional to the square of the speed.

▨ The thinking distance is affected by the condition of the driver (drink and/or drugs) and by distractions within the vehicle; the braking

▶▶

distance by the mechanical condition of the brakes and tyres and the type of road surface.

- Impact forces in accidents are reduced by the use of seat belts and air bags by passengers, and by the incorporation of crumple zones in the vehicle design.
- The Global Positioning System (GPS) depends on the reception, by an in-car computer, of signals from four or more satellites orbiting the Earth. The computer works out the position of the car using *trilateration*: it solves equations for its *x*-, *y*-, *z*- and *t*-co-ordinates from information about the distance of the car from the satellites.

Exam-style Questions

1 In a driving manual, it is suggested that, when driving at $13\,m\,s^{-1}$ (about $45\,km$ per hour), a driver should always keep a minimum of two car-lengths between the driver's car and the one in front.
   (a) Suggest a scientific justification for this safety tip, making reasonable assumptions about the magnitudes of any quantities you need.
   (b) How would you expect the length of this 'exclusion zone' to depend on speed, for speeds higher than $13\,m\,s^{-1}$?

2 A student, standing on the platform at a railway station, notices that the first two carriages of an arriving train pass her in $2.0\,s$, and the next two in $2.4\,s$. The train is decelerating uniformly. Each carriage is $20\,m$ long. When the train stops, the student is opposite the last carriage. How many carriages are there in the train?

3 A ball is to be kicked so that, at the highest point of its path, it just clears a horizontal cross-bar on a pair of goal-posts. The ground is level and the cross-bar is $2.5\,m$ high. The ball is kicked from ground level with an initial speed of $8.0\,m\,s^{-1}$.

(a) Calculate the angle of projection of the ball and the distance of the point where the ball was kicked from the goal-line.
(b) Also calculate the horizontal velocity of the ball as it passes over the cross-bar.
(c) For how long is the ball in the air before it reaches the ground on the far side of the cross-bar?

4 An athlete competing in the long jump leaves the ground at an angle of $28°$ and makes a jump of $7.40\,m$.
   (a) Calculate the speed at which the athlete took off.
   (b) If the athlete had been able to increase this speed by 5%, what percentage difference would this have made to the length of the jump?

5 A hunter, armed with a bow and arrow, takes direct aim at a monkey hanging from the branch of a tree. At the instant that the hunter releases the arrow, the monkey takes avoiding action by releasing its hold on the branch. By setting up the relevant equations for the motion of the monkey and the motion of the arrow, show that the monkey was mistaken in its strategy.

3. Work, energy and power

Machines such as wind turbines do work for us. They change energy from one form into some other more useful form. This chapter deals with the subject of energy in its various forms. Not only is the availability of useful forms of energy important, but also the rate at which it can be converted from one form to another. The rate of converting energy or using energy is known as power.

At the end of Section 3.1 you should be able to:
- show an understanding of the concept of, and define, work done by a force in terms of the product of the force and displacement in the direction of the force, and define the joule
- calculate the work done in a number of situations, including the work done by a gas which is expanding against a constant external pressure: $W = p\Delta V$
- select and use the equation for density: $\rho = m/V$
- select and use the equation for pressure: $p = F/A$, where F is the force normal to the area A
- derive, from definitions of pressure and density, the equation $p = \rho g h$.

The word 'work' is in use in everyday English language but it has a variety of meanings. In physics, the word **work** has a definite meaning. The vagueness of the term 'work' in everyday speech causes problems for some students when they come to give a precise scientific definition of work.

'I'm going to work today.'

'Where do you work?'

'I've done some work in the garden.'

'Lots of work was done lifting the box.'

'I've done my homework.'

> Work is done when a force moves the point at which it acts (the point of application) in the direction of the force.

work done = force × distance moved by force in the direction of the force

It is very important to include direction in the definition of work done. A car can be pushed horizontally quite easily but, if the car is to be lifted off its wheels, much more work has to be done and a machine, such as a car-jack, is used.

When a force moves its point of application in the direction of the force, the force does work and the work done *by* the force is said to be *positive*.

Figure 3.1 The weight-lifter uses a lot of energy to lift the weights but they can be rolled along the ground with little effort.

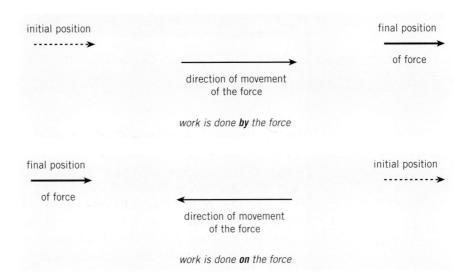

initial position

direction of movement of the force

final position

of force

*work is done **by** the force*

final position

of force

direction of movement of the force

initial position

*work is done **on** the force*

Figure 3.2

Conversely, if the direction of the force is *opposite* to the direction of movement, work is done *on* the force. This work done is then said to be *negative*. This is illustrated in Figure 3.2.

An alternative name for distance moved in a particular direction is **displacement**. Displacement is a vector quantity, as is force. However, work done has no direction, only size, and is a scalar quantity. It is measured in joules (J).

> When a force of one newton moves its point of application by one metre in the direction of the force, one joule of work is done.

work done in joules = force in newtons × distance moved in metres in the direction of the force

It follows that a joule (J) may be said to be a newton metre (N m). If the force and the displacement are not both in the same direction, then the component of the force in the direction of the displacement must be found. Consider a force F acting along a line at an angle θ to the displacement, as shown in Figure 3.4. The component of the force along the direction of the displacement is $F \cos \theta$.

work done for displacement $x = F \cos \theta \times x$
$= Fx \cos \theta$

Figure 3.3 The useful work done by the small tug-boat is found using the component of the tension in the rope along the direction of motion of the ship.

Note that the component $F \sin \theta$ of the force is at right angles to the displacement. Since there is no displacement in the direction of this component, no work is done.

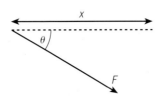

Figure 3.4

Example

A child tows a toy by means of a string as shown in Figure 3.5.

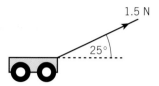

Figure 3.5

The tension in the string is 1.5 N and the string makes an angle of 25° with the horizontal. Calculate the work done in moving the toy horizontally through a distance of 265 cm.

$$work\ done = horizontal\ component\ of\ tension \times distance\ moved$$
$$= 1.5\cos 25 \times \frac{265}{100}$$
$$= \textbf{3.6 J}$$

Now it's your turn

1 A box weighs 45 N. Calculate the work done in lifting the box through a vertical height of:
   (a) 4.0 m,
   (b) 67 cm.

2 A force of 36 N acts at an angle of 55° to the vertical. The force moves its point of application by 64 cm in the direction of the force. Calculate the work done by:
   (a) the horizontal component of the force,
   (b) the vertical component of the force.

Density and pressure

You will be familiar with the terms density and pressure from your earlier work. We will summarise their definitions again, and bring them together to see an important link between them.

The density of a substance is defined as its mass per unit volume.

$$\rho = m/V$$

The symbol for density is ρ (Greek rho) and its SI unit is kg m^{-3}.

Example

An iron sphere of radius 0.18 m has mass 190 kg. Calculate the density of iron.

First calculate the volume of the sphere from $V = \frac{1}{4}\pi r^3$. This works out at 0.024 m^3. Application of the formula for density gives $\rho = $ **7800 kg m^{-3}**.

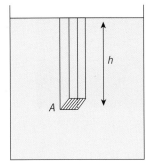

Figure 3.6 Column of liquid above the area A

Pressure is defined as force per unit area, where the force F acts perpendicularly to the area A.

$$p = F/A$$

The symbol for pressure is p and its SI unit is the pascal (Pa), which is equal to one newton per square metre (N m^{-2}).

The link between pressure and density comes when we deal with liquids, or with fluids in general. Consider a point at a depth h below the surface of a liquid in a container. What is the pressure due to the liquid? Very simply, the pressure is caused by the weight of the column of liquid above a small area at that depth, as shown in Figure 3.6. The weight of the column is $W = mg = \rho A h g$, and the pressure is $W/A = \rho g h$.

$$p = \rho g h$$

The pressure is proportional to the depth below the surface of the liquid. If an external pressure, such as atmospheric pressure, acts on the surface of the liquid, this must be taken into account in calculating the absolute pressure. The absolute pressure is the sum of the external pressure and the pressure due to the depth below the surface of the liquid.

Example

Calculate the excess of pressure over atmospheric at a point 1.2 m below the surface of the water in a swimming pool. The density of water is 1.0×10^3 kg m^{-3}.

This is a straightforward calculation from $p = \rho g h$. Substituting, $p = 1.0 \times 10^3 \times 9.8 \times 1.2 = $ **1.2 × 10^4 Pa**.

If the total pressure had been required, this value would be added to atmospheric pressure p_A. Taking p_A to be 1.01×10^5 Pa, the total pressure is 1.13×10^5 Pa.

Now it's your turn

Calculate the difference in blood pressure between the top of the head and the soles of the feet of a student 1.3 m tall, standing upright. Take the density of blood as 1.1×10^3 kg m^{-3}.

Work done by an expanding gas

A building can be demolished with explosives (Figure 3.7). When the explosives are detonated, large quantities of gas at high pressure are produced. As the gas expands, it does work by breaking down the masonry. In this section, we will derive an equation for the work done when a gas changes its volume.

Figure 3.7 Explosives produce large quantities of high-pressure gas. When the gas expands, it does work in demolishing the building.

Consider a gas contained in a cylinder by means of a frictionless piston of area A, as shown in Figure 3.8. The pressure p of the gas in the cylinder is equal to the atmospheric pressure outside the cylinder. This pressure may be thought to be constant.

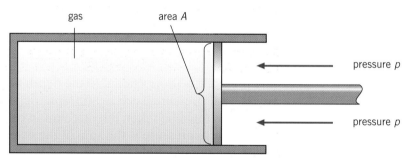

Figure 3.8

Since $pressure = \dfrac{force}{area}$, the gas produces a force F on the piston given by

$$F = pA$$

When the gas expands at constant pressure, the piston moves outwards through a distance x. So,

work done by the gas = force × distance moved
$$W = pAx$$

However, Ax is the increase in volume of the gas ΔV. Hence,

$$W = p\Delta V$$

When the volume of a gas changes at constant pressure,

work done = pressure × change in volume

When the gas *expands*, work is done *by* the gas. If the gas *contracts*, then work is done *on* the gas.

Figure 3.9 It is expanding gases pushing on the pistons which cause work to be done by the engine in a car.

Remember that the unit of work done is the joule (J). The pressure must be in pascals (Pa) or newtons per metre squared (N m^{-2}) and the change in volume in metres cubed (m^3).

Example

A sample of gas has a volume of 750 cm^3. The gas expands at a constant pressure of 1.4×10^5 Pa so that its volume becomes 900 cm^3. Calculate the work done by the gas during the expansion.

$$\text{change in volume } \Delta V = (900 - 750)$$
$$= 150 \text{ cm}^3$$
$$= 150 \times 10^{-6} \text{ m}^3$$

$$\text{work done by gas} = p\Delta V$$
$$= (1.4 \times 10^5) \times (150 \times 10^{-6})$$
$$= \textbf{21 J}$$

Now it's your turn

1 The volume of air in a tyre is 9.0×10^{-3} m^3. Atmospheric pressure is 1.0×10^5 Pa. Calculate the work done against the atmosphere by the air when the tyre bursts and the air expands to a volume of 2.7×10^{-2} m^3.

2 High-pressure gas in a spray-can has a volume of 250 cm^3. The gas escapes into the atmosphere through a nozzle, so that its final volume is four times the volume of the can. Calculate the work done by the gas, given that atmospheric pressure is 1.0×10^5 Pa.

Section 3.1 Summary

- When a force moves its point of application in the direction of the force, work is done.
- *Work done = Fx cos θ*, where *θ* is the angle between the direction of the force *F* and the displacement *x*.
- When a force of one newton moves its point of application by one metre in the direction of the force, one joule of work is done.
- *Pressure = the force normal to an area divided by that area*, or $p = F/A$
- When a gas expands at constant pressure:
  work done = pressure × change in volume, or $W = p\Delta V$
- Density = mass/volume, or $\rho = m/V$
- The pressure *p* at a depth *h* below the surface of a liquid of density *ρ* is $p = \rho gh$

Section 3.1 Questions

1 A force *F* moves its point of application by a distance *x* in a direction making an angle *θ* with the direction of the force, as shown in Figure 3.10.

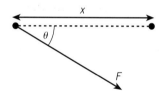

Figure 3.10

The force does an amount *W* of work. Copy and complete the following table.

F/N	x/m	θ/°	W/J
15	6.0	0	
15	6.0	90	
15	6.0	30	
46		23	6.4
2.4×10^3	1.6×10^2		3.1×10^5
	2.8	13	7.1×10^3

2 An elastic band is stretched so that its length increases by 2.4 cm. The force required to stretch the band increases linearly from 6.3 N to 9.5 N. Calculate:
(a) the average force required to stretch the elastic band,
(b) the work done in stretching the band.

3 When water boils at an atmospheric pressure of 101 kPa, 1.00 cm³ of liquid becomes 1560 cm³ of steam. Calculate the work done against the atmosphere when a saucepan containing 550 cm³ of water is allowed to boil dry.

3.2 Energy

At the end of Section 3.2 you should be able to:

- distinguish between gravitational potential energy, electric potential energy and elastic potential energy
- derive, from the defining equation $W = Fs$, the formula $\Delta E_p = mg\Delta h$ for potential energy changes near the Earth's surface
- recall and use the formula $\Delta E_p = mg\Delta h$ for potential energy changes near the Earth's surface
- show an understanding and use the relationship between force and potential energy in a uniform field to solve problems
- derive, from the equations of motion, the formula $E_k = \frac{1}{2}mv^2$
- recall and apply the formula $E_k = \frac{1}{2}mv^2$
- describe examples of energy in different forms, its conversion and conservation, and apply the principle of energy conservation (the work-energy principle) to simple examples and to solve problems
- apply the principle of conservation of energy to determine the speed of an object falling in the Earth's gravitational field
- show an appreciation for the implications of energy losses in practical devices and use the concept of efficiency to solve problems

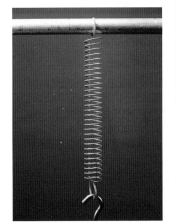

Figure 3.11 The spring stores energy as it is stretched, releasing the energy as it returns to its original shape.

In order to wind up a spring, work has to be done because a force must be moved through a distance. When the spring is released, it can do work; for example, making a child's toy move. When the spring is wound, it stores the ability to do work.

The ability to do work is called **energy**.

Since work done is measured in joules (J), energy is also measured in joules. Table 3.1 lists some typical values of energy.

Table 3.1 Typical energy values

	order of magnitude of energy/J
radioactive decay of a nucleus	10^{-13}
sound of speech on ear for 1 second	10^{-8}
moonlight on face for 1 second	10^{-3}
beat of the heart	1
burning a match	10^3
large cream cake	10^6
energy released from 100 kg of coal	10^{10}
earthquake	10^{19}
energy received on Earth from the Sun in one year	10^{25}
rotational energy of the Milky Way Galaxy	10^{50}
estimated energy of formation of the Universe	10^{70}

Potential energy

Potential energy is the ability of an object to do work as a result of its position or shape.

We have already seen that a wound-up spring stores energy. This energy is potential energy because the spring is strained. More specifically, the energy may be called **elastic** (or **strain**) **potential energy**. Elastic potential energy is stored in objects which have had their shape changed elastically. Examples include stretched wires, twisted elastic bands and compressed gases.

Another form of potential energy is **electric potential energy**. The law of charges – like charges repel, unlike charges attract – means that work has to be done when charges are moved relative to one another. If, for example, two positive charges are moved closer together, work is done and the electric potential energy of the charges increases. The electric potential energy stored is released when the charges move apart. Conversely, if a positive charge moves closer to a negative charge, energy is released because there is a force of attraction.

Newton's law of gravitation tells us that all masses attract one another. We rely on the force of gravity to keep us on Earth! When two masses are pulled apart, work is done on them and so they gain **gravitational potential energy**. If the masses move closer together, they lose gravitational potential energy.

Changes in gravitational potential energy are of particular importance for an object near to the Earth's surface because we frequently do work raising masses and, conversely, the energy stored is released when the mass is lowered again. An object of mass m near the Earth's surface has weight mg, where g is the acceleration of free fall. This weight is the force with which the Earth attracts the mass (and the mass attracts the Earth).

If the mass moves a *vertical* distance Δh,

$$work\ done = force \times distance\ moved$$
$$= mg\Delta h$$

Figure 3.12 The cars on the rollercoaster have stored gravitational potential energy. This energy is released as the cars fall.

When the mass is raised, the work done is stored as *gravitational potential energy* and this energy can be recovered when the mass falls.

Change in gravitational potential energy $\Delta E_p = mg\Delta h$

It is important to remember that, for the energy to be measured in joules, the mass m must be in kilograms, the acceleration g in metres second^{-2} and the change in height Δh in metres.

Notice that a zero point of gravitational potential energy has not been stated. We are concerned with *changes* in potential energy when a mass rises or falls.

Example

A shop assistant stacks a shelf with 25 tins of beans, each of mass 460 g. Each tin has to be raised through a distance of 1.8 m. Calculate the gravitational potential energy gained by the tins of beans, given that the acceleration of free fall is 9.8 m s^{-2}.

total mass raised = 25 × 460 = 11 500 g
$$= 11.5 \text{ kg}$$

increase in potential energy = $m \times g \times \Delta h$
$$= 11.5 \times 9.8 \times 1.8$$
$$= \textbf{202.86 J}$$

Now it's your turn

The acceleration of free fall is 9.8 m s^{-2}. Calculate the change in gravitational potential energy when:
(a) a person of mass 70 kg climbs a cliff of height 19 m,
(b) a book of mass 940 g is raised vertically through a distance of 130 cm,
(c) an aircraft of total mass 2.5 × 10^5 kg descends by 980 m.

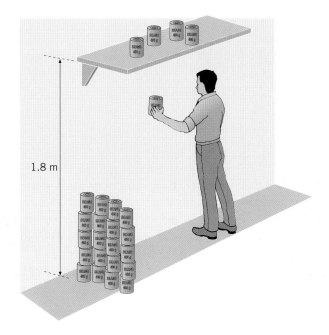

1.8 m

Figure 3.13

Kinetic energy

As an object falls, it loses gravitational potential energy and in so doing it speeds up. Energy is associated with a moving object. In fact, we know that a moving object can be made to do work as it slows down. For example, a moving hammer hits a nail and, as it stops, does work to drive the nail into a piece of wood.

Kinetic energy is energy due to motion.

Figure 3.14 When the mass falls, it gains kinetic energy and drives the pile into the ground.

Consider an object of mass m moving with a constant acceleration a. In a distance s, the object accelerates from velocity u to velocity v. Then, by referring to the equations of motion (see Chapter 2),

$$v^2 = u^2 + 2as$$

By Newton's law (see Chapter 4), the force F giving rise to the acceleration a is given by

$$F = ma$$

Combining these two equations,

$$v^2 = u^2 + 2\frac{F}{m}s$$

Re-arranging,

$$mv^2 = mu^2 + 2Fs$$
$$2Fs = mv^2 - mu^2$$
$$Fs = \tfrac{1}{2}mv^2 - \tfrac{1}{2}mu^2$$

By definition, the term Fs is the work done by the force moving a distance s. Therefore, since Fs represents work done, then the other terms in the equation, $\tfrac{1}{2}mv^2$ and $\tfrac{1}{2}mu^2$, must also have the units of work done, or energy (see Chapter 1). The magnitude of each of these terms depends on velocity squared and so $\tfrac{1}{2}mv^2$ and $\tfrac{1}{2}mu^2$ are terms representing energy which depends on velocity (or speed). The kinetic energy E_k of an object of mass m moving with speed v is given by

$$E_k = \tfrac{1}{2}mv^2$$

For the kinetic energy to be in joules, mass must be in kilograms and speed in metres per second.

The full name for the term $E_k = \tfrac{1}{2}mv^2$ is *translational kinetic energy* because it is energy due to an object moving in a straight line. It should be remembered that rotating objects also have kinetic energy and this form of energy is known as rotational kinetic energy.

Here we have used an important concept: that work done is equal to the transfer of energy. We showed that the work done (the Fs term on the left-hand side of the equation above) is equal to the increase of kinetic energy (the $\tfrac{1}{2}mv^2 - \tfrac{1}{2}mu^2$ term on the right-hand side). This idea is sometimes called the **work-energy principle**.

Example

Calculate the kinetic energy of a car of mass 900 kg moving at a speed of 20 m s^{-1}. State the form of energy from which the kinetic energy is derived.

$$kinetic\ energy = \tfrac{1}{2}mv^2$$
$$= \tfrac{1}{2} \times 900 \times 20^2$$
$$= \mathbf{1.8 \times 10^5\ J}$$

This energy is derived from the chemical energy of the fuel.

Now it's your turn

1 Calculate the kinetic energy of a car of mass 800 kg moving at 100 kilometres per hour.

2 A cycle and cyclist have a combined mass of 80 kg and are moving at 5.0 m s^{-1}. Calculate:
  (a) the kinetic energy of the cycle and cyclist,
  (b) the increase in kinetic energy for an increase in speed of 5.0 m s^{-1}.

Energy conversion and conservation

Newspapers sometimes refer to a 'Global Energy Crisis'. In the near future, there may well be a shortage of fossil fuels. Fossil fuels are sources of chemical energy. It would be more accurate to refer to a 'Fuel Crisis'. When chemical energy is used, the energy is transformed into other forms of energy, some of which are useful and some of which are not. Eventually, all the chemical energy is likely to end up as energy that is no longer useful to us. For example, when petrol is burned in a car engine, some of the chemical energy is converted into the kinetic energy of the car and some is wasted as heat (thermal) energy. When the car stops, its kinetic energy is converted into heat energy in the brakes. The outcome is that the chemical energy has been converted into heat energy which dissipates in the atmosphere and is of no further use. However, the total energy present in the Universe has remained constant. All energy changes are governed by the **law of conservation of energy**. This law states that

Energy cannot be created or destroyed. It can only be converted from one form to another.

There are many different forms of energy and you will meet a number of these during your Physics studies. Some of the more common forms are listed in Table 3.2.

Table 3.2 Forms of energy

energy	notes
potential energy	energy due to position
kinetic energy	energy due to motion
elastic or strain energy	energy due to stretching an object
electrical energy	energy associated with moving electric charge
sound energy	a mixture of potential and kinetic energy of the particles in the wave
wind energy	a particular type of kinetic energy
light energy	energy of electromagnetic waves
solar energy	light energy from the Sun
chemical energy	energy released during chemical reactions
nuclear energy	energy associated with particles in the nuclei of atoms
thermal energy	sometimes called heat energy

We can use the idea of energy conversion to solve a practical example. Suppose that you accidentally drop a cup from the breakfast table. You know from experience that the cup will gather speed and may shatter when it reaches the tiled floor. In the physical analysis of the situation, the cup possessed potential energy due to its position when it was on the table. As soon as it is dropped, the potential energy is converted to kinetic energy. Can we calculate its speed when it reaches the floor?

For an object of mass m falling from rest from a height Δh, the initial potential energy is $mg\Delta h$. When it has fallen through this height, all its potential energy has been converted to kinetic energy. Suppose that the speed of the object is v: its kinetic energy is then $\frac{1}{2}mv^2$. Applying the idea that energy must be conserved, we know that the change in potential energy is equal to the change in kinetic energy; thus

$$mg\Delta h = \frac{1}{2}mv^2$$

Cancelling m, we have

$$v^2 = 2g\Delta h$$

or

$$v = \sqrt{(2g\Delta h)}$$

This is a useful expression relating speed v to the height Δh through which an object falls in the Earth's gravitational field; you should try to remember it.

Example

Map out the energy changes taking place when a battery is connected to a lamp.

Chemical energy in battery → electrical energy in wires
→ light energy and heat energy in lamp

Now it's your turn

Map out the following energy changes:
(a) a child swinging on a swing,
(b) an aerosol can producing hair spray,
(c) a lump of clay thrown into the air which subsequently hits the ground.

Efficiency

In most energy changes some energy is 'lost' as heat (thermal) energy. For example, when a ball rolls down a slope, the total change in gravitational potential energy is not equal to the gain in kinetic energy because heat (thermal) energy has been produced as a result of frictional forces.

Efficiency gives a measure of how much of the total energy may be used and is not 'lost'.

$$efficiency = \frac{useful\ work\ done}{total\ energy\ input}$$

Efficiency may be given either as a ratio or as a percentage. Since energy cannot be created, efficiency can never be greater than 100% and a 'perpetual motion' machine is not possible (Figure 3.15).

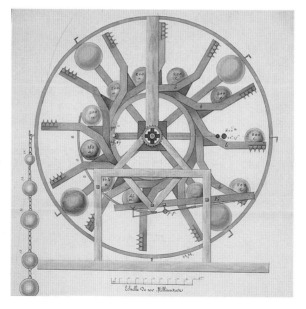

Figure 3.15 An attempt to design a machine to get something for nothing by breaking the law of conservation of energy.

A way of representing energy transfers pictorially is to use a *Sankey diagram*. This is a diagram which, in our case, represents the flow of energy through a conversion process, and shows what happens to the energy. (Sankey diagrams are also used to represent the flow of money, or materials, in relevant conversion processes.) Figure 3.16 is an example of a Sankey diagram for the very simple case of a filament electric light bulb.

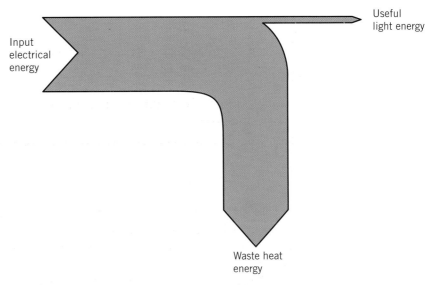

Input electrical energy

Useful light energy

Waste heat energy

Figure 3.16 Sankey diagram for a filament electric light bulb

Here the input energy is electrical, and is shown as a broad arrow at the left-hand side of the diagram. The useful output energy is light, and the 'lost' energy is heat. For the case of a filament lamp, it is often stated that only about 10% of the input energy appears as light; the remaining 90% is wasted heat. The feature of a Sankey diagram is that the widths of the arrows give a measure of the flow-rates. So here the width of the light output arrow is only a tenth of the width of the electrical energy input arrow. Note that the useful output continues the horizontal arrow of the input; the 'lost' energy is shown as an arrow (or arrows, in more complex situations) turning off to the vertical. These vertical arrows can be shown at the particular part of the process at which they occur. This presentation gives the viewer an immediate interpretation of what the important energy losses may be, and how important they are. Here it is very obvious that the 90% of the input energy that is turned into heat represents a serious loss. The traditional type of electric lamp is very inefficient.

Example

1 A man lifts a weight of 480 N through a vertical distance of 3.5 m using a rope and some pulleys. The man pulls on the rope with a force of 200 N and a length of 10.5 m of rope passes through his hands. Calculate the efficiency of the pulley system.

work done by man = force × distance moved
$$= 200 \times 10.5$$
$$= 2100 \, J$$

work done lifting load = 480×3.5
$$= 1680\,\text{J}$$
efficiency = work got out/work put in
$$= 1680/2100$$
$$= \textbf{0.80 or 80\%}$$

2 A manufacturer claims that a new design of energy-saving light bulb is 'Eight times more efficient than a traditional filament bulb'. Draw a Sankey diagram to show the claimed energy conversion process in the bulb.

If the new bulb is indeed eight times more efficient than the traditional bulb, the efficiency of which was assumed in the text above to be only 10%, the new bulb has an efficiency of 80%. That is, the useful light output is 80% of the electrical energy input. The energy 'lost' as heat is then 20% of the input. In the Sankey diagram (Figure 3.17) the width of the input arrow is 20 mm, and that of the light energy output is 16 mm. The width of the heat energy arrow is 4 mm.

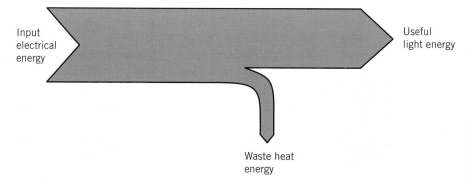

Figure 3.17

The manufacturer's claim is probably not accurate. One of the factors not taken into account is the point that the spectra of the energy-saving bulb and the traditional bulb are likely to be different, so that the light energy outputs have maxima at different wavelengths. The human eye responds to different wavelengths with different sensitivities. The comparison of the efficiencies of light sources, in terms of the reaction of the human eye, is a complex calculation.

Now it's your turn

1 An electric heater converts electrical energy into heat energy. Suggest why this process may be 100% efficient.

2 The electric motor of an elevator (lift) uses 630 kJ of electrical energy when raising the elevator and passengers, of total weight 12 500 N, through a vertical height of 29 m. Calculate the efficiency of the elevator.

Section 3.2 Summary

- Energy is needed to do work; energy is the ability to do work.
- Potential energy is the energy stored in a body due to its position or shape; examples are elastic potential energy, electric potential energy and gravitational potential energy.
- When an object of mass m moves vertically through a distance Δh, then the change in gravitational potential energy is given by:

$$\Delta E_p = mg\Delta h$$

  where g is the acceleration of free fall.
- Kinetic energy is the energy stored in a body due to its motion.
- For an object of mass m moving with speed v, the kinetic energy is given by: $E_k = \frac{1}{2}mv^2$
- Energy cannot be created or destroyed. It can only be converted from one form to another.
- *Efficiency = useful work done/total energy input*
- Energy flow in a process can be represented by a Sankey diagram.

Section 3.2 Questions

1 Name each of the following types of energy:
   (a) energy used in muscles,
   (b) energy stored in the Sun,
   (c) energy of water in a mountain lake,
   (d) energy captured by a wind turbine,
   (e) energy produced when a firework explodes,
   (f) energy of a compressed gas.

2 A child of mass 35 kg moves down a sloping path on a skate board. The sloping path makes an angle of 4.5° with the horizontal. The constant speed of the child along the path is $6.5\,\mathrm{m\,s^{-1}}$. Calculate:
   (a) the vertical distance through which the child moves in 1.0 s,
   (b) the rate at which potential energy is being lost ($g = 10\,\mathrm{m\,s^{-2}}$).

3 A stone of mass 120 g is dropped down a well. The surface of the water in the well is 9.5 m below ground level. The acceleration of free fall of the stone is $9.8\,\mathrm{m\,s^{-2}}$. Calculate, for the stone falling from ground level to the water surface:
   (a) the loss of potential energy,
   (b) its speed as it hits the water, assuming all the potential energy has been converted into kinetic energy.

4 An aircraft of mass 3.2×10^5 kg accelerates along a runway. Calculate the change in kinetic energy, in MJ, when the aircraft accelerates:
   (a) from zero to $10\,\mathrm{m\,s^{-1}}$,
   (b) from $30\,\mathrm{m\,s^{-1}}$ to $40\,\mathrm{m\,s^{-1}}$,
   (c) from $60\,\mathrm{m\,s^{-1}}$ to $70\,\mathrm{m\,s^{-1}}$.

5 In order to strengthen her legs, an athlete steps up on to a box and then down again 30 times per minute. The girl has mass 50 kg and the box is 35 cm high. The exercise lasts 4.0 minutes and as a result of the exercise, her leg muscles generate 120 kJ of heat energy. Calculate the efficiency of the leg muscles ($g = 10\,\mathrm{m\,s^{-2}}$).

6 By accident, the door of a refrigerator is left open. Use the law of conservation of energy to explain whether the temperature of the room will rise, stay constant or fall after the refrigerator has been working for a few hours.

3.3 The deformation of solids

At the end of Section 3.3 you should be able to:
- appreciate that deformation is caused by a force and that, in one dimension, the deformation can be tensile or compressive
- describe the behaviour of springs in terms of force, extension, elastic limit, Hooke's law and the spring constant (force per unit extension)
- distinguish between elastic and plastic deformation of a material
- deduce the strain energy in a deformed material from the area under the force–extension graph
- define and use the terms stress, strain, the Young modulus and ultimate tensile stress
- describe an experiment to determine the Young modulus of a metal in the form of a wire
- demonstrate knowledge of the force–extension graphs for typical ductile, brittle and polymeric materials, including an understanding of ultimate tensile stress.

When forces are applied to a solid body, its shape changes. The change may be very small, but nevertheless the forces affect the spacing of the atoms in the solid to a tiny extent and its external dimensions change. This change of shape is called **deformation**. If you think of a metal cylinder, the application of forces along the axis of the cylinder can either stretch it, making it longer, or squash it slightly, making it shorter. We call the deformation produced by these forces a **tensile** deformation for the stretching case, or a **compressive** deformation for the squashing case.

Hooke's law

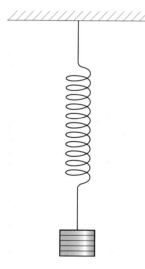

Figure 3.18 A loaded helical spring

A helical spring, attached to a fixed point, hangs vertically and has weights attached to its lower end, as shown in Figure 3.18. As the size of the weight is increased the spring becomes longer. The increase in length of the spring is called the **extension** of the spring and the weight attached to the spring is called the **load**.

If the load is increased greatly, the spring will permanently change its shape. However, for small loads, when the load is removed, the spring returns to its original length. The spring is said to have undergone an **elastic change**.

In an elastic change, a body returns to its original shape and size when the load on it is removed.

Figure 3.19 shows the variation with load of the extension of the spring. The section of the line from the origin to the point L is straight. In this region, the spring behaves elastically, and returns to its original length when the load is removed. The point L is referred to as the **elastic limit**. Beyond the point L, the spring is deformed permanently and the change is said to be **plastic**.

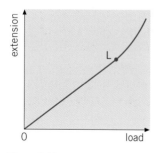

Figure 3.19 Extension of a loaded spring

The fact that there is a straight line relationship between load and extension for the elastic change is expressed in Hooke's law. It should be appreciated that, although we have used a spring as an illustration, the law applies to any object, provided the elastic limit has not been exceeded.

Hooke's law states that, provided the elastic limit is not exceeded, the extension of a body is proportional to the applied load.

The law can be expressed in the form of an equation

force $F \propto$ extension ΔL

Removing the proportionality sign gives

$F = k\Delta L$

where k is a constant, known as the **force constant, elastic constant** or **spring constant**.

The force constant (spring constant) is the force per unit extension.

The unit of the constant is newton per metre ($N\,m^{-1}$).

Example

An elastic cord has an unextended length of 25 cm. When the cord is extended by applying a force at each end, the length of the cord becomes 40 cm for forces of 0.75 N. Calculate the force constant of the cord.

extension of cord = 15 cm
force constant = 0.75/0.15 (extension in metres)
= **5.0 N m^{-1}**

Now it's your turn

1 Calculate the force constant for a spring which extends by a distance of 3.5 cm when a load of 14 N is hung from its end.

2 A steel wire extends by 1.5 mm when it is under a tension of 45 N. Calculate:
   (a) the force constant of the wire,
   (b) the tension required to produce an extension of 1.8 mm, assuming that the elastic limit is not exceeded.

Strain energy

When an object has its shape changed by forces acting on it, the object is said to be **strained**. Work has to be done by the forces to cause this strain. Provided that the elastic limit is not exceeded, the object can do work as it returns to its original shape when the forces are removed. Energy is stored in the body as potential energy when it is strained. This particular form of potential energy is called elastic potential energy or strain potential energy, or simply **strain energy**.

Strain energy is energy stored in a body due to change of shape.

Consider the spring shown in Figure 3.18. To produce an extension x, the force applied at the lower end of the spring increases linearly with extension from zero to a value F. The average force is $\frac{1}{2}F$ and the work done W by the force is therefore

$W = \text{average force} \times \text{extension}$ \quad (see Section 3.1)

$\quad = \frac{1}{2}Fx$

However, the force constant k is given by the equation

$F = k\Delta L$

The value ΔL is equivalent to x. Therefore, substituting for F,

$\text{strain energy } W = \frac{1}{2}k(\Delta L)^2$

The energy is in joules if k is in newtons per metre and x (equal to ΔL) is in metres.

A graph of load (y-axis) against extension (x-axis) enables strain energy to be found even when the graph is not linear (Figure 3.20). We have shown that strain energy is given by

$\text{strain energy } W = \frac{1}{2}F\Delta L$

$\quad = \frac{1}{2}Fx$

$\quad = \frac{1}{2}kx^2$

The expression $\frac{1}{2}Fx$ represents the area between the straight line on Figure 3.20 and the x-axis. This means that strain energy is represented by the area under the line on a graph of load (y-axis) plotted against extension (x-axis).

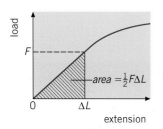

Figure 3.20 Strain energy is given by the area under the graph.

Example

A spring has a force constant $65\,\text{N m}^{-1}$ and is extended elastically by $1.2\,\text{cm}$. Calculate the strain energy stored in the spring.

$\text{strain energy } W = \frac{1}{2}kx^2$

$\quad = \frac{1}{2} \times 65 \times (1.2 \times 10^{-2})^2$

$\quad = \textbf{4.7} \times \textbf{10}^{-3}\,\textbf{J}$

Now it's your turn

1 A wire has a force constant of $5.5 \times 10^4\,\text{N m}^{-1}$. It is extended elastically by $1.4\,\text{mm}$. Calculate the strain energy stored in the wire.

2 A rubber band has a force constant of $180\,\text{N m}^{-1}$. The work done in extending the band is $0.16\,\text{J}$. Calculate the extension of the band.

The Young modulus

The difficulty with using the force constant is that the constant is different for each specimen of a material having a different shape. It would be far more convenient if we had a constant for a particular material which would enable us to find extensions knowing the constant and the dimensions of the specimen. This is possible using the **Young modulus**.

We have already mentioned the term **strain**. When an object of original length L is extended by an amount ΔL, the strain (ε) is defined as

$$strain = \frac{extension}{original\ length}$$

$$\varepsilon = \Delta L/L$$

Strain is the ratio of two lengths and does not have a unit.

The strain produced within an object is caused by a **stress**. In our case, we are dealing with changes in length and so the stress is referred to as a **tensile stress**. When a tensile force F acts normally to an area A, the stress (σ) is given by

$$stress = \frac{force}{area\ normal\ to\ the\ force}$$

$$\sigma = F/A$$

The unit of tensile stress is newtons per metre squared ($N\ m^{-2}$). This unit is also the unit of pressure and so an alternative unit for stress is the pascal (Pa).

In Figure 3.20, we plotted a graph of load against extension. Since load is related to stress and extension is related to strain, a graph of stress plotted against strain would have the same basic shape, as shown in Figure 3.21. Once again, there is a straight line region between the origin and L, the elastic limit. In this region, changes of strain with stress are elastic. In the region where the changes are elastic, it can be seen that

$$stress \propto strain$$

or, removing the proportionality sign,

$$stress = E \times strain$$

The constant E is known as the Young modulus of the material.

$$Young\ modulus\ E = \frac{stress}{strain}$$

The unit of the Young modulus is the same as that for stress because strain is a ratio and has no unit. Some values of the Young modulus for different materials are shown in Table 3.3.

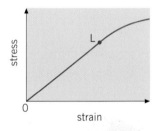

Figure 3.21 Stress–strain graph

Table 3.3 Young modulus for different materials

material	Young modulus E/Pa
aluminium	0.70×10^{11}
copper	1.1×10^{11}
steel	2.1×10^{11}
glass	0.41×10^{11}

The Young modulus of a metal in the form of a wire may be measured by applying loads to a wire and measuring the extensions caused. The original length and the cross-sectional area must also be measured. A suitable laboratory arrangement is shown in Figure 3.22. A copper wire is often used.

This is because, for wires of the same diameter under the same load, a copper wire will give larger, more measurable, extensions than a steel wire. (Why is this?) A paper flag with a reference mark on it is attached to the wire at a distance of just less than one metre from the clamped end. The original length L is measured from the clamped end to the reference mark, using a metre rule. The diameter d of the wire is measured using a micrometer screw gauge, and the cross-sectional area A calculated from $A = \frac{1}{4}\pi d^2$. Extensions ΔL are measured as masses m are added to the mass-carrier. (Think of a suitable way of measuring these extensions.) Be careful not to exceed the linear, Hooke's law, region. The load F is calculated from $F = mg$. A graph of ΔL against F has gradient L/EA, so the Young modulus E is equal to $L(A \times \text{gradient})$.

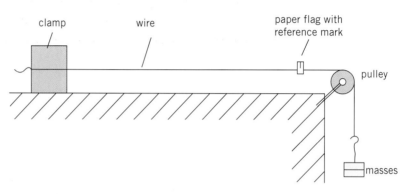

Figure 3.22 Simple experiment to measure the Young modulus of a wire

Example

A steel wire of diameter 1.0 mm and length 2.5 m is suspended from a fixed point and a mass of weight 45 N is suspended from its free end. The Young modulus of the material of the wire is 2.1×10^{11} Pa. Assuming that the elastic limit of the wire is not exceeded, calculate:
(a) the applied stress,
(b) the strain,
(c) the extension of the wire.

(a) $\text{area} = \pi \times (0.5 \times 10^{-3})^2$
        $= 7.9 \times 10^{-7}\,\text{m}^2$
  $\text{stress} = \text{force/area}$
        $= 45/7.9 \times 10^{-7}$
        $= \mathbf{5.7 \times 10^7\,Pa}$

(b) $\text{strain} = \text{stress/Young modulus}$
        $= 5.7 \times 10^7/2.1 \times 10^{11}$
        $= \mathbf{2.7 \times 10^{-4}}$

(c) $\text{extension} = \text{strain} \times \text{length}$
        $= 2.7 \times 10^{-4} \times 2.5$
        $= 6.8 \times 10^{-4}\,\text{m}$
        $= \mathbf{0.68\,mm}$

THE HENLEY COLLEGE LIBRARY

Now it's your turn

1 A copper wire of diameter 1.78 mm and length 1.4 m is suspended from a fixed point and a mass of weight 32 N is suspended from its free end. The Young modulus of the material of the wire is 1.1×10^{11} Pa. Assuming that the elastic limit of the wire is not exceeded, calculate:
   (a) the applied stress,
   (b) the strain,
   (c) the extension of the wire.

2 An elastic band of area of cross-section 2.0 mm² has an unextended length of 8.0 cm. When stretched by a force of 0.4 N, its length becomes 8.3 cm. Calculate the Young modulus of the elastic.

The behaviour of different materials under tensile stress

The force–extension graphs of all metals in the form of wires have the general shape shown in Figure 3.23: a straight line portion through the origin (the Young modulus is measured in this linear region), and then a region in which the extension increases more rapidly than the force. Eventually, well into the region of plastic deformation, application of larger and larger forces will cause the cross-section of the wire to form a narrow neck, so that the extension continues to increase without the addition of further force. The wire eventually breaks.

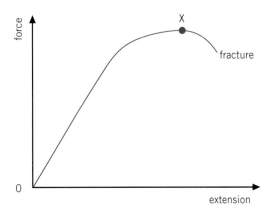

Figure 3.23 Force–extension graph for a ductile material

From the force at which the curve is a maximum (point X), a quantity known as the **ultimate tensile stress** can be calculated. This is the maximum force divided by the original cross-sectional area of the wire (not the area of the narrow neck, which would be almost impossible to specify). The ultimate tensile stress gives an idea of the maximum stress that the wire could support; note that this is *not* the same as the stress when the wire finally breaks. (Note also that the terminology 'breaking point' is not satisfactory, as it is not clear to what the 'point' refers; is it a position on the wire, the length at which the wire breaks, the force required to break the wire, the breaking stress or the breaking strain?)

Figure 3.23 is characteristic of materials which can be drawn out into wires, or **ductile** materials. Ductility is a characteristic of many metals.

Another class of material shows the behaviour sketched in Figure 3.24. A glass fibre is a typical example. The fibre extends elastically with a linear relation between force and extension over an appreciable range. Very soon after the limit of proportionality, the fibre snaps. This is called brittle fracture, and many amorphous substances like glass are classified as **brittle** materials. Their characteristic is that deformation obeys Hooke's law over practically the whole range of extensions; there is very little, if any, plastic deformation. This distinction tells us why, when a glass beaker is dropped on the floor, it shatters into many pieces. The glass is incapable of plastic deformation, and if the impact causes a deformation taking it out of the elastic region, it has to break. A metal beaker, on the other hand, is deformed plastically by the impact, ending up with a permanent deformation in the form of a dent, but not breaking.

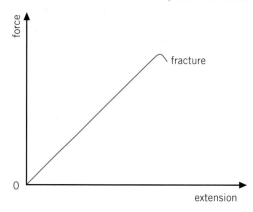

Figure 3.24 Force–extension graph for a brittle material

A typical polymeric material (such as rubber) has another entirely different force–extension graph. This is sketched in Figure 3.25 for a rubber cord. Only a very small part of this curve, near the origin, is sufficiently linear to use it to calculate the Young modulus. An important point to note is that the cord can be stretched to many times its original length before it breaks; the curve also shows a very extensive region in which the cord will return to its original length when the stretching force is removed. However, this may not be a case of simple elastic extension because, although the cord is not permanently deformed, it may not return to its original length along the same path. This is called **elastic hysteresis**. How is energy conserved? The graphs show that the strain energy required to deform the rubber (the area under the graph – see Figure 3.20) is greater than the work done by the

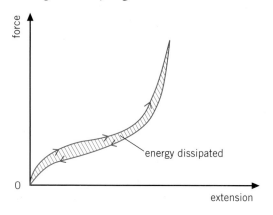

Figure 3.25 Force–extension graph for a polymeric material showing elastic hysteresis

material in returning to its original length. The excess energy, represented by the area bounded by the two curves, must be energy dissipated within the rubber, showing up as an increase in temperature. Types of rubber with large elastic hysteresis are used in vibration absorbers for mounting oscillating machinery and preventing the vibrations being transmitted to the floor. The mechanical energy of the oscillations is converted to thermal energy in the mounts.

Section 3.3 Summary

- Forces on an object can cause tensile deformation (stretching) or compressive deformation (squashing).
- An elastic change occurs when an object returns to its original shape and size when the load is removed from it.
- Hooke's law states that, for an elastic change, extension is proportional to load.
- The force constant (elastic constant, spring constant) k is the ratio of force to extension.
- *Strain energy* $= \frac{1}{2}kx^2$
- *Tensile strain = extension/original length*
- *Tensile stress = force/cross-sectional area*
- *Young modulus = stress/strain*
- Ductile and brittle materials have characteristic force–extension graphs.
- Polymeric materials such as rubber typically show elastic hysteresis.

Section 3.3 Questions

1 A spring has an unextended length of 12.4 cm. When a load of 4.5 N is suspended from the spring, its length becomes 13.3 cm. Calculate:
(a) the force constant of the spring,
(b) the length of the spring for a load of 3.5 N.

2 The elastic cord of a catapult has an force constant of 700 N m^{-1}. Calculate the energy stored in the elastic cord when it is extended by 15 cm.

3 Two wires each have length 1.8 m and diameter 1.2 mm. One wire has a Young modulus of 1.1×10^{11} Pa and the other 2.2×10^{11} Pa.

One end of each wire is attached to the same fixed point and the other end of each wire is attached to the same load of 75 N so that each has the same extension. Assuming that the elastic limit of the wires is not exceeded, calculate the extension of the wires.

4 Distinguish between ductile and brittle materials, making reference to the force–extension graph for each type of material. Name an example of each type.

5 Explain what is meant by elastic hysteresis.

3.4 Power

At the end of Section 3.4 you should be able to:
- define power as work done per unit time and derive power as the product of force and velocity.

We have seen that energy is the ability to do work. Consider a family car and a grand prix racing car which both contain the same amount of fuel. They are capable of doing the same amount of work, but the racing car is able to travel much faster. This is because the engine of the racing car can convert the chemical energy of the fuel into useful energy at a much faster rate. The engine is said to be more powerful. **Power** is the rate of doing work. Power is given by the formula

$$power = \frac{work\ done}{time\ taken}$$

The unit of power is the watt (symbol W) and is equal to a rate of working of 1 joule per second. This means that a light bulb of power 1 W will convert 1 J of electrical energy to other forms of energy (e.g. light and heat) every second. Some typical values of power are shown in Table 3.4.

Table 3.4 Values of power

	power/W
power to operate a small calculator	50×10^{-6}
light power from a torch	4×10^{-3}
loudspeaker output	5
manual labourer working continuously	100
water buffalo working continuously	750
hair dryer	3×10^3
motor car engine	80×10^3
electric train	5×10^6
electricity generating station output	2×10^9

Power, like energy, is a scalar quantity.

Care must be taken when referring to power. It is common in everyday language to say that a strong person is 'powerful'. In physics, strength, or force, and power are *not* the same. Large forces may be exerted without any movement and thus no work is done and the power is zero! For example, a large rock resting on the ground is not moving, yet it is exerting a large force.

Consider a force F which moves a distance x at constant speed v in the direction of the force, in time t. Referring back to Section 3.1, the work done W by the force is given by

$$W = Fx$$

Dividing both sides of this equation by time t gives

$$\frac{W}{t} = F\frac{x}{t}$$

Now, $\dfrac{W}{t}$ is the rate of doing work, i.e. the power P and $\dfrac{x}{t} = v$. Hence,

$P = Fv$

power = force × speed

Example

A small electric motor is used to lift a weight of 1.5 N through a vertical distance of 120 cm in 2.7 s. Calculate the useful power output of the motor.

work done = force × distance moved
  = 1.5 × 1.2 (the distance must be in metres)
  = 1.8 J

  power = work done/time taken
    = (1.8/2.7)
    *= **0.67 W***

Now it's your turn

1 Calculate the electrical energy converted into thermal energy when an electric fire, rated at 2.4 kW, is left switched on for a time of 3.0 minutes.

2 The output power of the electric motors of a train is 3.6 MW when the train is travelling at $30\,m\,s^{-1}$. Calculate the total force opposing the motion of the train.

3 A boy of mass 60 kg runs up a flight of steps in a time of 1.8 s. There are 22 steps and each one is of height 20 cm. Calculate the useful power developed in the boy's legs. (The acceleration of free fall is $10\,m\,s^{-2}$.)

The kilowatt hour

Every household has to pay the 'electricity bill'. Electricity is vital in modern living and this energy does not come free of charge. It is important to realise that what is paid for is electrical energy, not electrical power. Since many electrical appliances in the home have a power of the order of kilowatts and we use them for hours, the joule, as a unit of energy, is too small. For example, an electric fire of power 3 kW, used for 2 hours would use

Figure 3.26 A domestic digital electricity meter

$3000 \times 2 \times 60 \times 60 = 21600000$ joules, i.e. 21.6 million joules of energy! Instead, in many parts of the world, electrical energy is purchased in kilowatt hours (kW h).

One kilowatt hour is the energy expended when work is done at the rate of 1 kilowatt for a time of 1 hour.

$$1\,\text{kW h} = 1.0\,\text{kW} \times 1\,\text{hour}$$
$$= 1000\,\text{W} \times 3600\,\text{s}$$
$$1\,\text{kW h} = 3.6 \times 10^6\,\text{J}$$

The kilowatt hour is sometimes referred to as the Unit of energy. Electricity meters in the home (Figure 3.26) are often shown as measuring Units, where 1 Unit = 1 kW h.

Example

Calculate the cost of using an electric fire, rated at 2.5 kW for a time of 6.0 hours, if 1 kW h of energy costs 7.0 pence.

energy used = $2.5 \times 6.0 = 15\,\text{kW h}$
cost = 15×7.0
= **105 pence**

Now it's your turn

1 A television set is rated at 280 W. Calculate the cost of watching a three-hour film if 1 kilowatt hour of electrical energy costs 8 pence.

2 An electric kettle is rated at 2.4 kW. Electrical energy costs 8 pence per kW h. The kettle takes 1.0 minute to boil sufficient water for two mugs of coffee. Calculate the cost of making this amount of coffee on three separate occasions.

3 Electrical energy generating companies sometimes measure their output in gigawatt years. Calculate the number of kilowatt hours in 6.0 gigawatt years.

Section 3.4 Summary

- Power is the rate of doing work.
- *Work done = power × time taken*
- The unit of power is the watt (W).
- 1 watt = 1 joule per second.
- *Power = force × speed*
- Electrical energy may be measured in kilowatt hours (kW h).
- 1 kW h is the energy expended when work is done at the rate of 1000 watts for a time of 1 hour.

Section 3.4 Questions

1 The lights in a school laboratory have a total power of 600 W and are left on for 7.0 hours each day. In order to reduce fuel bills, it is decided to have the lights switched on only when there are people in the laboratory. This amounts to a total time of 4.5 hours per day. Assuming that the laboratory is used for 200 days each year, calculate the saving, if 1 kW h of energy costs 7.0 pence.

2 A car travelling at speed v along a horizontal road moves against a resistive force F given by the equation

$$F = 400 + kv^2$$

where F is in newtons, v in m s^{-1} and k is a constant.

At speed $v = 15\,\text{m s}^{-1}$, the resistive force F is 1100 N.
(a) Calculate, for this car:
  (i) the power necessary to maintain the speed of $15\,\text{m s}^{-1}$,
  (ii) the total resistive force at a speed of $30\,\text{m s}^{-1}$,
  (iii) the power required to maintain the speed of $30\,\text{m s}^{-1}$.
(b) Determine the energy expended in travelling 1.2 km at a constant speed of:
  (i) $15\,\text{m s}^{-1}$,
  (ii) $30\,\text{m s}^{-1}$.
(c) Using your answers to (b), suggest why, during a fuel shortage, the maximum permitted speed of cars may be reduced.

3.5 Moment of a force

At the end of Section 3.5 you should be able to:
- define and apply the moment of a force
- show an understanding that a couple is a pair of forces which tends to produce rotation only
- define and apply the torque of a couple
- show an understanding that, when there is no resultant force and no resultant torque, a system is in equilibrium
- apply the principle of moments to solve problems
- show an understanding that the weight of a body may be taken as acting at a single point known as its centre of gravity
- describe a simple experiment to determine the centre of gravity of an object.

When a force acts on an object, the force may cause the object to move in a straight line. It could also cause the object to spin (rotate).

Think about a metre rule held in the hand at one end so that the rule is horizontal (Figure 3.27). If a weight is hung from the ruler we can feel a

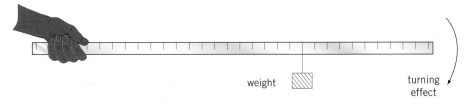

weight   turning effect

Figure 3.27 Turning effect on a metre rule

turning effect on the ruler. The turning effect increases if the weight is increased or it is moved further from the hand along the ruler. The turning effect acts at the hand where the metre rule is pivoted. Keeping the weight and its distance along the rule constant, the turning effect can be changed by holding the ruler at an angle to the horizontal. The turning effect becomes smaller as the rule approaches the vertical position.

The turning effect of a force is called the **moment** of the force.

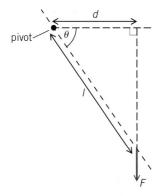

The moment of a force depends on the magnitude of the force and also on the distance of the force from the pivot or fulcrum. This distance must be defined precisely. In the simple experiment above, we saw that the moment of the force depended on the angle of the ruler to the horizontal. Varying this angle meant that the line of action of the force from the pivot varied (Figure 3.28). The distance required when finding the moment of a force is the perpendicular distance d of the line of action of the force from the pivot.

> The moment of a force is defined as the product of the force and the perpendicular distance of the line of action of the force from the pivot.

Figure 3.28 Finding the moment of a force

Referring to Figure 3.28, the force has magnitude F and acts at a point distance l from the pivot. Then, when the ruler is at angle θ to the horizontal,

$$moment\ of\ force = F \times d$$
$$= F \times l \cos \theta$$

Since force is measured in newtons and distance is measured in metres, the unit of the moment of a force is newton metre (N m).

Example

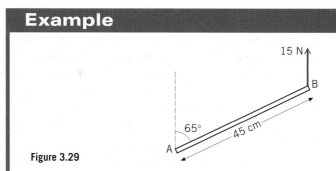

Figure 3.29

In Figure 3.29, a light rod AB of length 45 cm is held at A so that the rod makes an angle of 65° to the vertical. A vertical force of 15 N acts on the rod at B. Calculate the moment of the force about the end A.

$$moment\ of\ force = force \times perpendicular\ distance\ from\ pivot$$
$$= 15 \times 0.45 \sin 65$$

(Remember that the distance must be in metres.)
$$= \textbf{6.1\,N\,m}$$

> **Now it's your turn**
>
> Referring to Figure 3.29, calculate the moment of the force about A for a vertical force of 25 N with the rod at an angle of 30° to the vertical.

Couples

When a screwdriver is used, we apply a turning effect to the handle. We do not apply one force to the handle because this would mean the screwdriver would move sideways. Rather, we apply two forces of equal size but opposite direction on opposite sides of the handle (Figure 3.30).

> A **couple** consists of two forces, equal in magnitude but opposite in direction whose lines of action do not coincide.

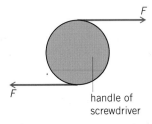

Figure 3.30 Two forces acting as a couple

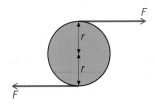

Figure 3.31 Torque of a couple

Consider the two parallel forces, each of magnitude F acting as shown in Figure 3.31 on opposite ends of a diameter of a disc of radius r. Each force produces a moment about the centre of the disc of magnitude Fr in a clockwise direction. The total moment about the centre is $F \times 2r$ or $F \times$ *perpendicular distance between the forces*.

Although a turning effect is produced, this turning effect is not called a moment because it is produced by two forces, not one. Instead, this turning effect is referred to as a **torque**. The unit of torque is the same as that of the moment of a force i.e. newton metre.

> The torque of a couple is the product of one of the forces and the perpendicular distance between the forces.

Figure 3.32 Tightening a wheel nut requires the application of a torque.

It is interesting to note that, in engineering, the tightness of nuts and bolts is often stated as the maximum torque to be used when screwing up the nut on the bolt. Spanners used for this purpose are called torque wrenches because they have a scale on them to indicate the torque that is being applied.

Example

Calculate the torque produced by two forces, each of magnitude 30 N, acting in opposite directions with their lines of action separated by a distance of 25 cm.

torque = force × separation of forces
            = 30 × 0.25   (distance in metres)
            = **7.5 N m**

Now it's your turn

The torque produced by a person using a screwdriver is 0.18 N m. This torque is applied to the handle of diameter 4.0 cm. Calculate the force applied to the handle.

The principle of moments

A metre rule may be balanced on a pivot so that the rule is horizontal. Hanging a weight on the rule will make the rule rotate about the pivot. Moving the weight to the other side of the pivot will make the rule rotate in the opposite direction. Hanging weights on both sides of the pivot as shown in Figure 3.33 means that the ruler may rotate clockwise, or anticlockwise or it may remain horizontal. In this horizontal position, there is no resultant turning effect and so the total turning effect of the forces in the clockwise direction equals the total turning effect in the anticlockwise direction. You can check this very easily with the apparatus of Figure 3.33.

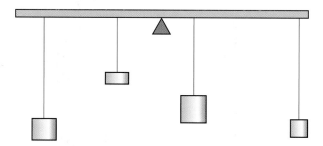

Figure 3.33

When a body has no tendency to change its speed of rotation, it is said to be in **rotational equilibrium**.

The **principle of moments** states that, for a body to be in rotational equilibrium, the sum of the clockwise moments about any point must equal the sum of the anticlockwise moments about that same point.

Examples

1 Some weights are hung from a light rod AB as shown in Figure 3.34. The rod is pivoted. Calculate the magnitude of the force *F* required to balance the rod horizontally.

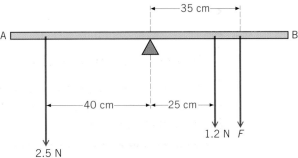

Figure 3.34

Sum of clockwise moments = $(0.25 \times 1.2) + 0.35F$
Sum of anticlockwise moments = 0.40×2.5
By the principle of moments
$$(0.25 \times 1.2) + 0.35F = 0.40 \times 2.5$$
$$0.35F = 1.0 - 0.3$$
$$F = \mathbf{2.0\,N}$$

2 A girl holds a stone of weight 44 N in the palm of her hand, as shown in Figure 3.35. Her forearm is held in the horizontal position by the vertical tension *T* in the biceps muscle (the muscle in the upper arm). The biceps muscle is attached to the forearm bone 5.0 cm from the elbow joint. The weight of her forearm and hand is 25 N with its centre

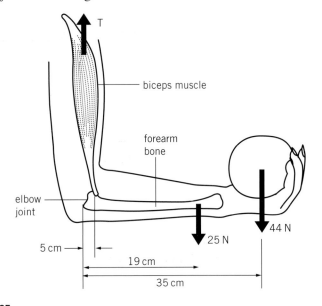

Figure 3.35

of gravity 19 cm from the elbow joint. The centre of gravity of the stone is 35 cm from the joint. Find the tension T in the biceps muscle.

It helps to draw a simple model of the arm, as in Figure 3.36.

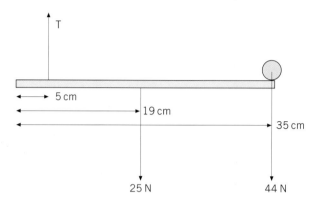

Figure 3.36

Sum of clockwise moments = $(0.19 \times 25) + (0.35 \times 44)$
Sum of anticlockwise moments = $0.05T$

By the principle of moments
$(0.19 \times 25) + (0.35 \times 44) = 0.05T$
$4.75 + 15.4 = 0.05T$
$T = 403\,\text{N}$

Now it's your turn

1 Some weights are hung from a light rod AB as shown in Figure 3.37. The rod is pivoted. Calculate the magnitude of the force F required to balance the rod horizontally.

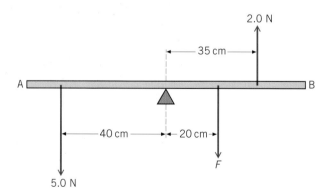

Figure 3.37

2   The girl in Example 2, still holding the stone, raises her forearm so
that it makes an angle of 30° with the horizontal, as shown in
Figure 3.38. The upper arm and biceps muscle remain vertical.

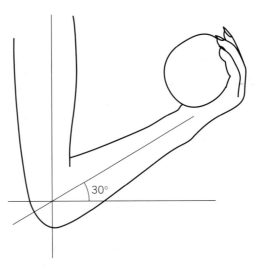

30°

Figure 3.38

Calculate the magnitude of the tension *T* in the biceps muscle
required to hold her arm stationary in this new position.

Centre of gravity

An object may be made to balance at a particular point. When it is balanced
at this point, the object does not turn and so all the weight on one side of the
pivot is balanced by the weight on the other side. Supporting the object at the
pivot means that the only force which has to be applied at the pivot is one to
stop the object falling – that is, a force equal to the weight of the object.
Although all parts of the object have weight, the whole weight of the object
appears to act at this balance point. This point is called the **centre of gravity**
(C.G.) of the object.

> The centre of gravity of an object is the point at which the whole weight
> of the object may be considered to act.

The weight of a body can be shown as a force acting vertically downwards at
the centre of gravity.
    The centre of gravity of a body of an object can be determined by a simple
experiment. If the body is suspended freely from a pivot point, it will move
and settle so that the line of action of its weight (which is the vertical line
passing through its centre of gravity), when extended upwards, passes

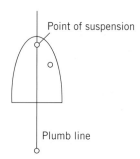

Point of suspension

Plumb line

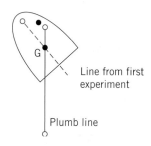

G

Line from first experiment

Plumb line

Figure 3.39

through the point of suspension. The experiment is often carried out with a two-dimensional body like a sheet of cardboard. The body is first allowed to come to rest, in equilibrium, when pivoted freely about a pin passing through a small hole in the card. Then a plumb line is attached to the pin to define the vertical. The vertical line passing through the point of suspension is traced onto the card. The process is then repeated with the body pivoted about a second point. The two lines which were traced to define the vertical at each pivot will intersect at a point which is the centre of gravity, as shown in Figure 3.39. The result can be checked by pivoting the card about a third point and again tracing the vertical; this line should pass through, or very close to, the intersection of the previous two lines.

For a uniform body such as a ruler, the centre of gravity is at the geometrical centre. For a circular piece of card, or a wheel, the centre of gravity lies at the centre of the circle; for a sphere, the centre of gravity is at the centre of the sphere. For some shapes of object, for example an L-shaped sheet of cardboard, the centre of gravity is not on the card at all, but in the space between the two arms of the L.

Equilibrium

The principle of moments gives the condition necessary for a body to be in rotational equilibrium. However, the body could still have a resultant force acting on it which would cause it to accelerate linearly. Thus, for complete equilibrium, there cannot be any resultant force in any direction.

For a body to be **in equilibrium**:

1 The sum of the forces in any direction must be zero,
2 The sum of the moments of the forces about any point must be zero.

The first condition for equilibrium can be written in a form which is more useful for solving problems. The condition refers to 'the sum of forces in any direction'. This emphasises the vector nature of a force. Remember that vectors can be resolved into two directions at right angles (see Section 1.2). If we label these directions as the x- and y-directions, we can say that the first condition means that the sum of the x-components of all the forces must be zero, and the sum of the y-components must be zero too. The first condition could thus be written as

1 The sums of the components of the forces in two perpendicular directions must each be zero.

Another way of stating the first condition is by drawing a vector diagram of the forces acting on the object. The forces are in equilibrium when the vectors representing them are joined up, nose to tail, and form a closed figure. For the simple case of three forces acting at a point on an object, this closed figure is a triangle. An experiment to demonstrate this triangle of forces was described in Section 1.2 (pages 13–14).

Section 3.5 Summary

- The moment of a force is a measure of the turning effect of the force.
- The moment of a force is the product of the force and the perpendicular distance of the line of action of the force from the pivot.
- A couple consists of two equal forces acting in opposite directions whose lines of action do not coincide.
- The torque of a couple is a measure of the turning effect of the couple.
- The torque of a couple is the product of one of the forces and the perpendicular distance between the lines of action of the forces.
- The principle of moments states that the sum of the clockwise moments about a point is equal to the sum of the anticlockwise moments about the point.
- The centre of gravity of a body is the point at which the whole weight of the body may be considered to act.
- For a body to be in equilibrium:
  1 the sum of the forces in any direction must be zero,
  2 the sum of the moments of the forces about any point must be zero.

Section 3.5 Questions

1 A uniform rod has a weight of 14 N. It is pivoted at one end and held in a horizontal position by a thread tied to its other end, as shown in Figure 3.40. The thread makes an angle of 50° with the horizontal. Calculate:
(a) the moment of the weight of the rod about the pivot,
(b) the tension T in the thread required to hold the rod horizontally.

2 A ruler is pivoted at its centre of gravity and weights are hung from the ruler as shown in Figure 3.41. Calculate:
(a) the total anticlockwise moment about the pivot,
(b) the magnitude of the force F.

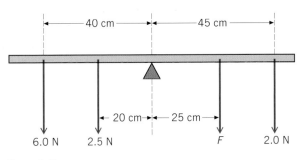

Figure 3.41

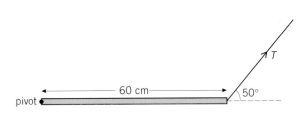

Figure 3.40

3 A uniform plank of weight 120 N rests on two stools as shown in Figure 3.42. A weight of 80 N is placed on the plank, midway between the stools. Calculate:
(a) the force acting on the stool at A,
(b) the force acting on the stool at B.

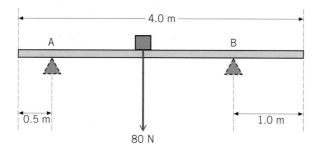

Figure 3.42

4 A nut is to be tightened to a torque of 16 N m. Calculate the force which must be applied to the end of a spanner of length 24 cm in order to produce this torque.

Exam-style Questions

1 (a) State what is meant by:
  (i) work done,
  (ii) power.
(b) (i) Show that, when a gas expands by an amount ΔV against a constant pressure p, the work done W is given by

$$W = p\Delta V$$

  (ii) Explain why, when the gas in (i) expands suddenly, it is likely to cool.
(c) The air in a car tyre has a volume of $7.8 \times 10^3 \, cm^3$. The tyre suddenly bursts and the air expands to 3.5 times its original volume in a time of 2.5 ms. Atmospheric pressure is $1.0 \times 10^5 \, Pa$. Calculate:
  (i) the work done by the air in the tyre during the expansion,
  (ii) the mean power dissipated in the tyre burst.

2 (a) Explain, by reference to two examples from different areas of physics, what is meant by *potential* energy.
(b) (i) State the law of conservation of energy.
  (ii) Explain how the law may be applied to a ball dropped from a height h on to a horizontal surface and which finally comes to rest on the surface after several bounces.

4. Force and motion

In Chapter 2 we described motion in terms of displacement, velocity and acceleration. Now we will try to explain *why* bodies move. We shall realise that a force is required to make a body accelerate. This will introduce Newton's three laws of motion, the fundamental laws which state and define what forces do in relation to motion.

4.1 Force

At the end of Sections 4.1–4.4 you should be able to:

■ show a qualitative understanding of frictional forces and viscous forces, including air resistance
■ state each of Newton's laws of motion
■ define the newton
■ recall and solve problems using the relationship $F = ma$ appreciating that acceleration and force are always in the same direction
■ recall that, according to the special theory of relativity, $F = ma$ cannot be used for a particle travelling at very high speeds, because its mass increases
■ show an understanding that mass is the property of a body which resists changes in motion
■ describe and use the concept of weight as the effect of a gravitational field on a mass
■ show an understanding of the effects of fields of force, as exemplified by the forces on mass and charge in uniform gravitational and electric fields respectively.

When you push a trolley in a supermarket or pull a case behind you at an airport, you are exerting a force. When you hammer in a nail, a force is being exerted. When you drop a book and it falls to the floor, the book is falling because of the force of gravity. When you lean against a wall or sit on a chair, you are exerting a force. Forces can change the shape or dimensions of bodies. You can crush a drinks can by squeezing it and applying a force; you can stretch a rubber band by pulling it. In everyday life, we have a good understanding of what is meant by force and the situations in which forces are involved. In physics the idea of force is used to add detail to the descriptions of moving objects.

As with all physical quantities, a method of measuring force must be established. One way of doing this is to make use of the fact that forces can change the dimensions of bodies in a reproducible way. It takes the same force to stretch a spring by the same amount (provided the spring is not overstretched by applying a very large force). This principle is used in the spring balance. A scale shows how much the spring has been extended, and the scale can be calibrated in terms of force. Laboratory spring balances are often called newton balances, because the newton is the SI unit of force.

Forces are vector quantities: they have magnitude as well as direction. Forces acting on a body are often shown by means of a force diagram drawn to scale, in which the forces are represented by lines of length proportional to the magnitude of the force, and in the appropriate direction (see Chapter 1).

4.2 Force and motion

The Greek philosopher Aristotle believed that the natural state of a body was a state of rest, and that a force was necessary to make it move and to keep it moving. In this argument, the greater the force, the greater the speed of the body.

Nearly two thousand years later, Galileo questioned this idea. He suggested that motion at a constant speed could be just as natural a state as the state of rest. He introduced an understanding of the effect of **friction** on motion.

Imagine a heavy box being pushed along a rough floor at constant speed (Figure 4.1). This may take a considerable force. The force required can be reduced if the floor is made smooth and polished, and reduced even more if a lubricant, for example grease, is applied between the box and the floor. We can imagine a situation where, when friction is reduced to a vanishingly small value, the force required to push the box at constant speed is also vanishingly small.

Galileo realised that the force of friction was a force that opposed the pushing force. When the box is moving at constant speed, the pushing force is exactly equal to the frictional force, but in the opposite direction, so that there is a net force of zero acting on the box. In the situation of vanishingly small friction, the box will continue to move with constant speed, because there is no force to slow it down.

Frictional forces are also important in considering the motion of a body through a liquid or gas. We use the term **viscous force** to describe the frictional force in a fluid (such as air resistance). The property of the fluid determining this force is the **viscosity** of the fluid. In Section 2.5 we considered the fact that parachutists eventually fall with a constant, terminal velocity because of air resistance.

The factors that determine the magnitude of the drag force are the viscosity of the fluid, the speed of the object as it passes through the fluid, and the frontal area of the object presented to the fluid. The drag force also depends on whether the object moves through the fluid slowly, in what is called streamline motion, or whether it is moving faster, so that the motion is turbulent. The speed at which turbulent flow kicks in is determined strongly by the shape of the object. At this level, you need only have a qualitative knowledge of the factors.

Figure 4.1

Newton's laws of motion

Figure 4.2 Isaac Newton

Isaac Newton (1642–1727) used Galileo's ideas to produce a theory of motion, expressed in his three laws of motion. The **first law of motion** re-states Galileo's deduction about the natural state of a body.

> Every body continues in its state of rest, or of uniform motion in a straight line, unless compelled to change that state by a net force.

This law tells us what a force does: it disturbs the state of rest or uniform motion of a body. The property of a body to stay in a state of rest or uniform motion is called **inertia**.

Newton's second law tells us what happens if a force is exerted on a body. It causes the velocity to change. A force exerted on a body at rest makes it move – it gives it a velocity. A force exerted on a moving body may make its speed increase or decrease, or change its direction of motion. A change in speed or velocity is an acceleration. Newton's second law defines the magnitude of this acceleration. It also introduces the idea of the mass of a body. Mass is a measure of the inertia of a body. The bigger the mass, the more difficult it is to change its state of rest or uniform motion. A simplified form of Newton's **second law** is

> For a body of constant mass, its acceleration is directly proportional to the net force applied to it.

The direction of the acceleration is in the direction of the net force. In a word equation, the law is

> *force = mass × acceleration*

and in symbols

> $F = ma$

where F is the force, m is the mass and a is the acceleration. Note that, according to the special theory of relativity, this equation cannot be used for a particle travelling at a very high speed (of the order of the speed of light). This is because, according to relativity theory, the mass depends on speed and increases as the speed v approaches the speed of light c.

In the equation $F = ma$ we have made the constant of proportionality equal to unity (that is, we use an equals sign rather than a proportionality sign) by choosing quantities with units which will give us this simple relation. In SI units, the force F is in newtons, the mass m in kilograms and the acceleration a in metres per second per second.

Figure 4.3

One newton is defined as the force which will give a mass of one kilogram an acceleration of one metre per second per second in the direction of the force.

When a force is applied to a body, that force is always applied by another body. When you push a supermarket trolley, the trolley experiences a force. That force has been applied by another body – you. Newton understood that the body on which the force is exerted applies another force back on the body which is applying the force. The supermarket trolley exerts a force on you as well. Newton's **third law** relates these two forces.

Whenever one body exerts a force on another, the second exerts an equal and opposite force of the same kind on the first.

Very often this law is stated as

To every action, there is an equal and opposite reaction.

But this statement does not highlight the very important point that the action force and the reaction force act on *different objects*. To take the example of the supermarket trolley, the action force exerted by you on the trolley is equal and opposite to the reaction force exerted by the trolley on you.

Newton's third law has applications in every branch of everyday life. We walk because of this law. When you take a step forward, your foot presses against the ground. The ground then exerts an equal and opposite force on you. This is the force, on you, which propels you in your path. Space rockets work because of the law (Figure 4.4). To expel the exhaust gases from the rocket, the rocket exerts a force on the gases. The gases exert an equal and opposite force on the rocket, sending it forward.

Figure 4.4 Space shuttle launch

Examples

1 An object of mass 1.5 kg is to be accelerated at 2.2 m s^{-2}. What force is required?

By Newton's second law, $F = ma = 1.5 \times 2.2 = \mathbf{3.3\,N}$.

2 A car of mass 1.5 tonnes (1.5×10^3 kg), travelling at 80 km h^{-1}, is to be stopped in 11 s. What force is required?

The acceleration of the car can be obtained from $v = u + at$ (see Chapter 3). The initial speed u is 80 km h^{-1}, or 22 m s^{-1}. The final speed v is 0. Then $a = -22/11 = -2.0$ m s^{-2}. This is negative because the car is decelerating.

By Newton's second law, $F = ma = 1.5 \times 10^3 \times 2.0 = \mathbf{3.0 \times 10^3\,N}$.

4.3 Weight

We saw in Chapter 2 that all objects released near the surface of the Earth fall with the same acceleration (the acceleration of free fall) if air resistance can be neglected. The force causing this acceleration is the gravitational attraction of the Earth on the object, or the force of gravity. The force of gravity which acts on an object is called the **weight** of the object. We can apply Newton's second law to the weight. For a body of mass m falling with the acceleration of free fall g, the weight W is given by

$$W = mg$$

The SI unit of force is the newton (N). This is also the unit of weight. The weight of an object is obtained from its mass in kilograms by multiplying the mass by the acceleration of free fall, 9.8 m s^{-2}. Thus a mass of one kilogram has a weight of 9.8 N. Because weight is a force and force is a vector, we ought to be aware of the direction of the weight of an object. It is towards the centre of the Earth. Because weight always has this direction, we do not need to specify direction every time we give the magnitude of the weight of objects.

The weight of a body is an example of the force acting on a mass in what is called a **field of force**. Near the surface of the Earth, this field is approximately constant and uniform. This means that in calculations we can take the same value of g, whatever the position on the surface of the Earth or for a short distance (compared with the Earth's radius) above it.

There are other sorts of fields of force. An important example is an electric field. An electric charge experiences a force in an electric field. There are significant similarities between the behaviour of a mass in a gravitational field and an electric charge in an electric field.

How do we measure mass and weight? If you hang an object from a newton balance, you are measuring its weight (Figure 4.5). The unknown weight of the object is balanced by a force provided by the spring in the balance. From a previous calibration, this force is related to the extension of the spring. There is the possibility of confusion here. Laboratory newton balances may, indeed, be calibrated in newtons. But all commercial spring

Figure 4.5 A newton balance

balances – for example, the balances at fruit and vegetable counters in supermarkets – are calibrated in kilograms. Such balances are really measuring the weight of the fruit and vegetables, but the scale reading is in mass units, because there is no distinction between mass and weight in everyday life. The average shopper thinks of 5 kg of apples as having a weight of 5 kg. In fact, the mass of 5 kg has a weight of 49 N.

The word 'balance' in the spring balance and in the laboratory top-pan balance relates to the balance of forces. In each case, the unknown force (the weight) is equalled by a force which is known through calibration.

A way of comparing masses is to use a beam balance, or lever balance Here the weight of the object is balanced against the weight of some masses, which have previously been calibrated in mass units. The word 'balance' here refers to the equilibrium of the beam: when the beam is horizontal the moment of the weight on one side of the pivot is equal and opposite to the moment on the other side of the pivot. Because weight is given by the product of mass and the acceleration of free fall, the equality of the weights means that the masses are also equal.

We have introduced the idea of weight by thinking about an object in free fall. But objects at rest also have weight: the gravitational attraction on a book is the same whether it is falling or whether it is resting on a table. The fact that the book is at rest tells us, by Newton's first law, that the resultant force acting on it is zero. So there must be another force acting on the book which exactly balances its weight. Here the table exerts an upwards force on the book. This force is equal in magnitude to the weight but opposite in direction. It is a **normal contact force**: 'contact' because it occurs due to the contact between book and table, and 'normal' because it acts perpendicularly to the plane of contact. The weight and normal contact forces are shown in Figure 4.6.

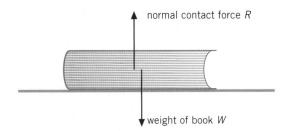
normal contact force R

weight of book W

Figure 4.6 A book resting on a table: forces on the book. (The forces act in the same vertical line, but are separated slightly for clarity.)

The book remains at rest on the table because the weight W of the book downwards is exactly balanced by the normal contact force R exerted by the table on the book. The vector sum of these forces is zero, so the book is in equilibrium. A very common mistake is to state that 'By Newton's third law, W is equal to R'. But these two forces are both acting on the book, and cannot be related by the third law. Third-law forces always act on *different* bodies. To see the application of the third law, think about the normal contact force R. This is an upwards force exerted by the *table*. The reaction to this is a downwards force R' exerted by the *book*. By Newton's third law, these forces are equal and opposite. This situation is illustrated in Figure 4.7.

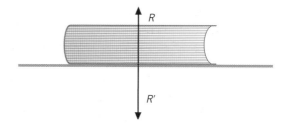

Figure 4.7 A book resting on a table: action and reaction forces, acting at the point of contact

Having considered the action and reaction forces between book and table, we ought to think about the reaction force to the weight of the book, regarded as an action force, even when the book is not on the table. This is not so easy, because there does not seem to be an obvious second force. But remember that the weight is due to the gravitational attraction of the Earth on the book. If the Earth attracts the book, the book also attracts the Earth. This gravitational force of the book on the Earth is the reaction force. We can test whether the two forces do, indeed, act on different bodies. The action force (the weight of the book) acts on the book. The reaction force (the attraction of the Earth to the book) acts on the Earth. Thus, the condition that action and reaction forces should act on different bodies is satisfied.

4.4 Problem solving

In dealing with problems involving Newton's laws, start by drawing a general sketch of the situation. Then consider each body in your sketch. Show all the forces acting on that body, both known forces and unknown forces you may be trying to find. Here it is a real help to try to draw the arrows which represent the forces in approximately the correct direction and approximately to scale. Label each force with its magnitude, or with a symbol if you do not know the magnitude. For each force, you must know the cause of the force (gravity, friction, and so on), and you must also know *on* what object that force acts and *by* what object it is exerted. This labelled diagram is referred to as a **free-body diagram**, because it detaches the body from the others in the situation. Having established all the forces acting on the body, you can use Newton's second law to find unknown quantities. This procedure is illustrated in the example which follows.

Newton's second law equates the resultant force acting on a body to the product of its mass and its acceleration. In some problems, the system of bodies is in equilibrium. They are at rest, or are moving in a straight line with uniform speed. In this case, the acceleration is zero, so the resultant force is also zero. In other cases, the resultant force is not zero and the objects in the system are accelerating.

For whichever case applies, you should remember that forces are vectors. You will probably have to resolve the forces into two components at right angles, and then apply the second law to each set of components separately. Problems can often be simplified by making a good choice of directions for resolution. You will end up with a set of equations, based on the application of Newton's second law, which must be solved to determine the unknown quantity.

Example

A box of mass 5.0 kg is pulled along a horizontal floor by a force P of 25 N, applied at an angle of 20° to the horizontal (Figure 4.8). A frictional force F of 20 N acts parallel to the floor.

Calculate the acceleration of the box.

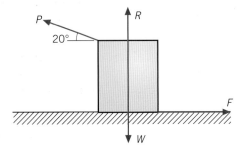

Figure 4.8

The free-body diagram is shown in Figure 4.8. Resolving the forces parallel to the floor, the component of the pulling force, acting to the left, is 25 cos 20 = 23.5 N. The frictional force, acting to the right, is 20 N. The resultant force to the left is thus 23.5 − 20.0 = 3.5 N.

By Newton's second law, $a = F/m = 3.5/5.0 = $ **0.70 m s^{-2}**.

Now it's your turn

Figure 4.9

A gardener pushes a lawnmower of mass 18 kg at constant speed. To do this requires a force P of 80 N directed along the handle, which is at an angle of 40° to the horizontal (Figure 4.9).
(a) Calculate the horizontal retarding force F on the mower.
(b) If this retarding force were constant, what force, applied along the handle, would accelerate the mower from rest to 1.2 m s^{-1} in 2.0 s?

Sections 4.1–4.4 Summary

- The force of friction opposes motion.
- Newton's laws of motion define *force*. The laws are:

  First law: Every body continues in its state of rest, or of uniform motion in a straight line, unless acted upon by a force.

  Second law: Force is proportional to mass × acceleration or $F = ma$, where force F is in newtons, mass m is in kilograms and acceleration a is in metres per second per second.

  Third law: Whenever one body exerts a force on another, the second exerts an equal and opposite force on the first.

- Weight W and mass m are related by:

  $W = mg$

  where g is the acceleration of free fall.

Sections 4.1–4.4 Questions

1 A net force of 95 N accelerates an object at 1.9 m s^{-2}. Calculate the mass of the object.

2 A parachute trainee jumps from a platform 3.0 m high. When he reaches the ground, he bends his knees to cushion the fall. His torso decelerates over a distance of 0.65 m. Calculate:
  (a) the speed of the trainee just before he reaches the ground,
  (b) the deceleration of his torso,
  (c) the average force exerted on his torso (of mass 45 kg) by his legs during the deceleration.

3 If the acceleration of a body is zero, does this mean that no forces act on it?

4 A railway engine pulls two carriages of equal mass with uniform acceleration. The tension in the coupling between the engine and the first carriage is T. Deduce the tension in the coupling between the first and second carriages.

5 What is your mass? What is your weight?

Exam-style Questions

1 A number of coplanar forces act on a mass that is free to move. Discuss fully the physical significance of the situation if the vector diagram representing the forces **(a)** is a closed polygon, **(b)** does not form a closed polygon.

2 A car has been driven into a field and has become stuck in the mud. Three light ropes are attached to a towing point at the front of the car. Forces parallel to the ground, and of the magnitudes shown, are applied to the car as shown in Figure 4.10.

Despite the efforts of the people pulling the ropes, the car remains stationary in the mud. Determine the magnitude and direction of the force, provided by the mud, which keeps the car stationary.

3 A truck, pulling a loaded trailer of total mass 300 kg, is moving along a level road at a speed of 12 m s^{-1}. The driver sees a child some distance ahead stepping into the road, and applies the brakes. The truck stops uniformly in a distance of 25 m. The trailer does not jack-knife during braking, but stays straight behind the truck. Calculate the horizontal force on the tow-hook while the truck is stopping.

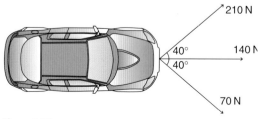

Figure 4.10

5. Electricity

Electricity is so vital to modern life that physicists and engineers need to understand it fully so that we can enjoy all its benefits. The aim of this chapter is to consider fundamental ideas about electric current and electric circuits in order to provide a basis for further study of electricity. We start by thinking about current as the flow of charge. We link the idea of potential difference to energy changes in a circuit. Resistance comes in as a measure of opposition to the flow of charge. The idea of resistance introduces Ohm's law. Kirchhoff's first and second laws are seen as the laws of conservation of charge and of energy, applied to simple electric circuits. Moving on to circuit theory, we shall derive the rules for combining resistors in series and in parallel, and also analyse complete circuits.

5.1 Charge and current

At the end of Section 5.1 you should be able to:
- relate an electric current to a flow of charged particles
- select and use the equation $\Delta Q = I\Delta t$
- define the *coulomb*
- distinguish between conventional current and electron flow
- state what is meant by *drift velocity*
- select and use the equation $I = nAve$
- distinguish between conductors, semiconductors and insulators.

All matter is made up of tiny particles called atoms, each consisting of a positively charged nucleus with negatively charged electrons moving around it.

Charge is measured in units called **coulombs** (symbol C). The charge on an electron is -1.6×10^{-19} C. Normally atoms have equal numbers of positive and negative charges, so that their overall charge is zero. But for some atoms it is relatively easy to remove an electron, leaving an atom with an unbalanced number of positive charges. This is called a **positive ion**.

Atoms in metals have one or more electrons which are not held tightly to the nucleus. These **free** (or mobile) **electrons** wander at random throughout the metal. But when a battery is connected across the ends of the metal, the free electrons drift towards the positive terminal of the battery, producing an **electric current**.

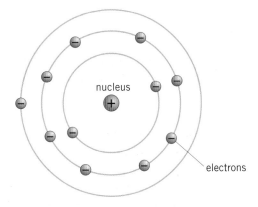

Figure 5.1 Atoms consist of a positively charged nucleus with negative electrons outside.

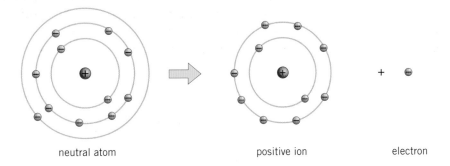

neutral atom   positive ion   electron

Figure 5.2 An atom with one or more electrons missing is a positive ion.

An electric current is a flow of electrons in a metal. The current could also be a flow of positive charge or of negative and positive charge. This can be shown as follows.

Two insulated metal plates are held vertically. The plates are connected in series with a high-voltage supply and a very sensitive ammeter as shown in Figure 5.3.

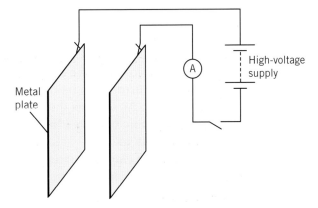

Figure 5.3 Demonstration of current due to positive and negative charge

When the switch is closed, there is no reading on the sensitive ammeter. A β-emitting radioactive source is then directed towards the space between the plates. Take care – there is a high voltage between the plates! A reading is seen on the ammeter.

The radioactive source ionises the air between the plates. The positive and negative ions move towards the plate that is oppositely charged. This movement of charge is an electric current.

The ionisation can also be produced using a candle flame. With the high-voltage supply switched off, a lighted candle is placed between the plates. When the supply is switched on, it is noted that the flame changes shape and there is a reading on the ammeter.

The migration of ions can be shown using the apparatus in Figure 5.4.

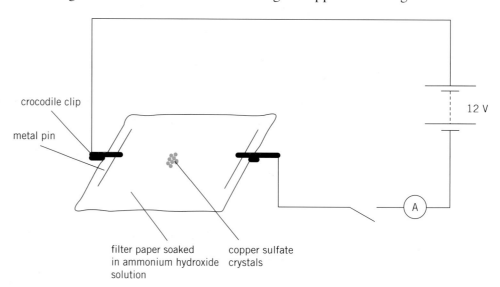

Figure 5.4 Migration of ions

The filter paper is soaked with ammonium hydroxide solution and is then connected into the circuit using crocodile clips and metal pins. Some copper sulfate crystals are placed near the centre of the filter paper. As soon as the switch is closed, the ammeter records a current. The blue coloured region (positive copper ions) moves slowly across the filter paper as the copper ions drift towards the negative connection.

The drift speed of the ions may be determined using a stopwatch and a ruler. It is found that the speed is relatively small (about $10^{-4}\,\mathrm{m\,s^{-1}}$). Although the current is registered at the instant that the switch is closed, the ions drift only slowly in the circuit.

An electric current can be modelled as the flow of water in a central heating system. As soon as the heating is switched on, there is a current of water in the whole system. However, the water itself moves relatively slowly through the system.

The size of an electric current is given by the rate of flow of charge, and is measured in units called amperes (or amps for short), with symbol A. A current of 3 amperes means that 3 coulombs pass a point in the circuit every second. In 5 seconds, a total charge of 15 coulombs will have passed the point. So,

charge = current × time

or

$$\Delta Q = I\Delta t$$

where the charge ΔQ is in coulombs when the current I is in amperes and the time Δt is in seconds. This gives a definition of the coulomb as

The **coulomb** is the charge passing a point in a circuit when there is a current of one ampere for one second.

Example

The current in the filament of a torch bulb is 0.03 A. How much charge flows through the bulb in 1 minute?

Using $\Delta Q = I\Delta t$, $\Delta Q = 0.03 \times 60$ (remember the time must be in seconds), so $\Delta Q = \textbf{1.8 C}$.

Now it's your turn

1 Calculate the current when a charge of 240 C passes a point in a circuit in a time of 2 minutes.

2 In a silver-plating experiment, 9.65×10^4 C of charge is needed to deposit a certain mass of silver. Calculate the time taken to deposit this mass of silver when the current is 0.20 A.

3 The current in a wire is 200 mA. Calculate:
   (a) the charge which passes a point in the wire in 5 minutes,
   (b) the number of electrons needed to carry this charge.
   (electron charge $e = -1.6 \times 10^{-19}$ C)

Conventional current

Early studies of the effects of electricity led scientists to believe that an electric current is the flow of 'something'. In order to develop a further understanding of electricity, they needed to know the direction of flow. It was decided that this flow in the circuit should be from the positive terminal of the battery to the negative. This current is called the **conventional current**, and is in the direction of flow of positive charge. We now know, in a metal, that the electric current is the flow of electrons in the opposite direction, from the negative terminal to the positive terminal. However, laws of electricity had become so firmly fixed in the minds of people that the idea of conventional current has persisted. But be warned! Occasionally we need to take into account the fact that electron flow is in the opposite direction to conventional current.

Current and drift velocity

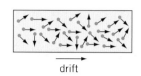

Figure 5.5 Drift velocity

The experiment outlined on page 96 showed that, when there is a current, the charge carriers drift through the conductor.

In many types of conductor, there are free charge carriers. In metals, these free charge carriers are electrons. When there is no current, the charge carriers move at high speeds (up to $10^6 \, \mathrm{m \, s^{-1}}$) and quite randomly in the conductor. Under the influence of a potential difference, there is an additional velocity along the conductor – the **drift velocity**. All the charge carriers move or drift through the conductor. Figure 5.5 illustrates electrons drifting through a metal.

The drift velocity of the charge carriers in a metal is of the order of $10^{-4} \, \mathrm{m \, s^{-1}}$.

Think about a sample of material of cross-sectional area A as shown in Figure 5.6.

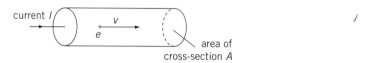

Figure 5.6 Charge carriers in a sample of a conductor

Under the influence of a potential difference, the charge carriers have a drift velocity v along the conductor.·

The current I in the conductor is equal to the rate of flow of charge through the end of the conductor (see page 96). The rate of flow of charge – the current – will depend on:

- the drift speed v. The greater the drift speed, the greater the current.
- the number of charge carriers per unit volume, i.e. the number density n. The greater the number density, the greater the current.
- the charge e on each carrier. This is taken as the elementary charge ($1.6 \times 10^{-19} \, \mathrm{C}$). Charge carriers (in particular, ions) may have a multiple of the elementary charge. The greater the charge, the greater the current.
- the area A of cross-section of the conductor. The greater the area, the greater the current.

All of these factors may be combined into one equation for the current in a conductor in terms of the drift velocity. That is

$$I = nAve$$

Conductors, semiconductors and insulators

Whether a material is classed as a conductor, an insulator or a semiconductor depends on the resistance of a cube of the material. This resistance will depend, in part, on the number density n of charge carriers in the material. This is summarised in Table 5.1.

Table 5.1 Number density of charge carriers in conductors, semiconductors and insulators

material	resistance	number density n
Conductors, e.g. metals	low	high, approximately 1 per atom, about $10^{29}\,m^{-3}$
Insulators, e.g. polythene, glass	very large	low, 10^6–$10^{12}\,m^{-3}$
Semiconductors, e.g. carbon, germanium	intermediate	very temperature dependent, increases as temperature rises 10^{20}–$10^{24}\,m^{-3}$

Section 5.1 Summary

▒ Electric current is the rate of flow of charge: $I = \Delta Q/\Delta t$.
▒ Conventional current is a flow of positive charge from positive to negative. In metals, current is carried by electrons, which travel from negative to positive.
▒ The number density of charge carriers in a material determines whether it is classed as a conductor, an insulator or a semiconductor.
▒ The current I in a conductor of cross-sectional area A is related to the drift speed v of charge carriers of number density n by the expression $I = nAve$, where e is the charge on the electron.

Section 5.1 Questions

1 A 240 V heater takes a current of 4.2 A. Calculate the charge that passes through the heater in 3 minutes.
2 Copper has one free electron per atom. The number density of atoms in a sample of copper is $2.1 \times 10^{29}\,m^{-3}$. The charge on the electron is 1.6×10^{-19} C. The current in a copper wire of diameter 1.2 mm is 7.2 A. Calculate the drift speed of free electrons in the wire.

5.2 Potential difference, resistance and heating

At the end of Section 5.2 you should be able to:
▒ define *potential difference* and the *volt*
▒ select and use the equation $W = VQ$
▒ define *resistance* and the *ohm*
▒ select and use the equation $R = V/I$
▒ define *power*

- select and use the equations $P = VI$, $P = I^2R$ and $P = V^2/R$
- select and use the equation *electrical energy = VIt*
- define and use the unit *kilowatt hour*
- explain how a fuse works as a safety device.

Potential difference

A cell makes one end of the circuit positive and the other negative. The cell is said to set up a **potential difference** across the circuit. Potential difference (p.d. for short) is measured in volts (symbol V), and is often called the voltage. You should never talk about the potential difference or voltage *through* a device, because it is in fact a difference *across* the ends of the device. The potential difference provides the energy to move charge through the device.

The potential difference between any two points in a circuit is a measure of the electrical energy transferred, or the work done, by each coulomb of charge as it moves from one point to the other. We already know that the unit of potential difference is the volt (V). Energy W is measured in joules, and charge Q in coulombs.

$$potential\ difference = \frac{energy\ transferred\ (or\ work\ done)}{charge}$$

or

$$V = \frac{W}{Q}$$

We can turn this relation round to get an expression for the electrical energy transferred or converted when a charge Q is moved through a potential difference V

$$energy\ transferred\ (work\ done) = potential\ difference \times charge$$

$$W = VQ$$

This relation gives a definition of the volt as

One **volt** is the potential difference between two points when one joule of energy is transferred by one coulomb passing from one point to the other.

In, Figure 5.7, one lamp is connected to a 240 V mains supply and the other to a 12 V car battery. Both lamps have the same current yet the 240 V lamp glows more brightly. This is because the energy supplied to each coulomb of charge in the 240 V lamp is 20 times greater than for the 12 V lamp.

Figure 5.7 A 240 V, 100 W mains lamp is much brighter than a 12 V, 5 W car light, but both have the same current. (**Do not try this experiment yourself** as it involves a large voltage.)

Example

Electrons in a particular television tube are accelerated by a potential difference of 20 kV between the filament and the screen. The charge of the electron is -1.6×10^{-19} C. Calculate the gain in kinetic energy of each electron.

Since $V = W/Q$, then $W = VQ$. The electrical energy transferred to the electron shows itself as the kinetic energy of the electron. Thus,

kinetic energy $= VQ = 20 \times 10^3 \times 1.6 \times 10^{-19}$
$$= \mathbf{3.2 \times 10^{-15} \, J}.$$
(Don't forget to turn the 20 kV into volts.)

Now it's your turn

1 An electron in a particle accelerator is said to have 1 million electronvolts of energy when it has been accelerated through a potential difference of 1 million volts. Calculate the energy, in joules, gained by the electron.

2 A torch bulb is rated 2.2 V, 0.25 A. Calculate:
   (a) the charge passing through the bulb in one second,
   (b) the energy transferred by the passage of each coulomb of charge.

Electrical power

Remember that power P is the rate of doing work, or of transferring energy. Remember also that $V = W/Q$. Divide each term on the right-hand side of this equation by time t, so that $V = (W/t)/(Q/t)$. W/t is power P, and Q/t is current I, so

$$potential\ difference = \frac{power}{current}$$

or

$$V = \frac{P}{I}$$

(*W/Q* is a quotient, so dividing *W* by *t* and dividing *Q* by *t* at the same time does not affect the final value because the *t* terms cancel each other out.)

The power is measured in watts (W) when the potential difference is in volts (V) and the current is in amperes (A). A voltmeter can measure the p.d. across a device and an ammeter the current through it; the equation above can then be used to calculate the power in the device.

Resistance

Connecting wires in circuits are often made from copper, because copper offers little opposition to the movement of electrons. The copper wire is said to have a low electrical resistance. In other words, copper is a good conductor.

Some materials, such as plastics, are poor conductors. These materials are said to be insulators, because under normal circumstances they conduct little or no current.

The **resistance** *R* of a wire is defined as the ratio of the potential difference *V* across the wire to the current *I* in it.

or

$$R = \frac{V}{I}$$

where the resistance is in **ohms** when the potential difference is in volts and the current in amperes. The symbol for ohms is the Greek capital letter omega, Ω. We have defined resistance in terms of a wire, but many devices have resistance. The general term for such a device is a **resistor**. (Note that the resistance of a resistor is measured in ohms, just as the volume of a tank is measured in m³. We do not refer to the 'm³' of a tank, nor to the 'ohms' of a resistor.)

The relation between resistance, potential difference and current means that, for a given potential difference, a high resistance means a small current, while a low resistance means a large current.

Example

The current in an electric immersion heater in a school experiment is 6.3 A when the p.d. across it is 12 V. Calculate the resistance of the heater.

Since *R* = *V*/*I*, the resistance *R* = 12/6.3 = **1.9 Ω**.

Now it's your turn

The current in a light-emitting diode is 20 mA when it has a potential difference of 2.0 V across it. Calculate its resistance.

Electrical heating

When an electric current passes through a resistor, it gets hot. This heating effect is sometimes called **Joule heating**. The electrical power P produced (dissipated) is given by $V = P/I$, which can be rearranged to give $P = VI$. We can obtain alternative expressions for power in terms of the resistance R of the resistor. Since $V = IR$, then

$$P = I^2R$$

and

$$P = \frac{V^2}{R}$$

For a given resistor, the power dissipated depends on the square of the current. This means that if the current is doubled, the power will be four times as great. Similarly, a doubling of voltage will increase the power by a factor of four.

Example

An electric immersion heater used in a school experiment has a current of 6.3 A when the p.d. across it is 12 V. Calculate the power of the heater.

Since $P = VI$, power = $12 \times 6.3 = $ **76 W**.

Now it's your turn

1 Show that a 100 W lamp connected to a mains supply of 240 V will have the same current as a 5 W car lamp connected to a 12 V battery. (See Figure 5.7)

2 An electric kettle has a power of 2.2 kW at 240 V. Calculate:
(a) the current in the kettle,
(b) the resistance of the kettle element.

Energy and the kilowatt hour

Since power is the rate at which work is done, then

energy or *work done = power × time*

If an electrical appliance is operated for a time t, then the electrical energy dissipated is given by

$$electrical\ energy = VIt = I^2Rt = \frac{V^2}{R} \times t$$

Example

An electric drill is rated as 750 W. During a day, the drill is used for a total time of 85 minutes. Calculate the energy dissipated in the drill.

Since *energy = power × time*,
$$energy = 750 \times 85 \times 60$$
$$= \textbf{3 825 000 J}$$

Now it's your turn

A kettle is rated at 1.4 kW. Calculate the length of time, in minutes, before the kettle has dissipated 1 176 000 J.

An electric drill and a kettle are common items in many homes. The examples above shown that even when these items are used for relatively short periods of time the numbers involved are very large if the energy is measured in joules.

In order to avoid such large numbers, the unit of electrical energy for domestic purposes (known as 'the unit') is the **kilowatt hour** (kW h).

The kilowatt hour is the amount of energy dissipated when energy is being transferred at a rate of 1.0 kW for 1 hour.

Since *energy = power × time*,
$$1.0\ kW\ h = 1000 \times 60 \times 60$$
$$= 3\ 600\ 000\ J$$
$$= 3.6\ MJ$$

Example

One unit of electrical energy costs 7 pence. A shower is rated as 6.5 kW. Calculate the cost of taking a shower lasting 5.0 minutes.

Energy used = 6.5 × 5/60 = 0.54 kW h
cost = 0.54 × 7 = **3.8 pence**

Now it's your turn

During the course of an evening, a household uses the following appliances for the times indicated.

Two 100 W lamps for 3 hours
Three 60 W lamps for 4 hours
Television, rated at 320 W, for 4 hours
Oven, rated at 3.5 kW, for 30 minutes
Electrical energy costs 6.5 pence per kW h. Calculate the cost of the energy used.

Fuses

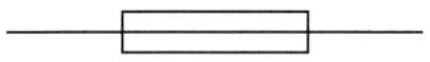

Figure 5.8 Circuit symbol for a fuse

A fuse is a safety device in an electrical circuit that cuts off the current to an appliance if the current becomes too large. A fuse consists of a short length (about 2 cm) of a metal wire that has a low melting point. The circuit symbol for a fuse is shown in Figure 5.8.

The fuse is connected in series with the supply to the circuit. In the case of a mains supply, it would be placed in the 'live' lead. When current passes through the fuse, it will be heated. If the current becomes too large, the fuse wire will melt and the current in the circuit will be cut off.

The diameter of the fuse wire and the material from which the wire is made will determine the maximum current in the fuse before it melts ('blows'). The rating of the fuse should be chosen so that appliances in the circuit will be protected from too large a current.

If the fuse is placed in the 'live' lead then, when the fuse melts, the appliance will be isolated from the mains supply and will thus be safe to touch.

One of the major causes of house fires is where an electrical circuit has too large a current.

Example

A television is rated as 320 W, 240 V. Fuses are available that are marked 1 A, 2 A, 5 A, 10 A and 13 A. Determine which fuse would be suitable for the television.

Since $P = VI$, current = 320/240 = 1.3 A.

The **2 A** fuse would be the most suitable one for use in the television.

Now it's your turn

An electric kettle is rated as 2.9 kW, 240 V.
What would be a suitable rating for a fuse that is to be used in the circuit for the kettle?

Section 5.2 Summary

- Potential difference (or voltage) measures the electrical energy transferred by each coulomb of charge: $V = W/Q$
- Resistance R of a resistor is defined as: $R = V/I$
- Electrical power $P = VI = I^2R = V^2/R$
- Electrical energy $= VIt = I^2Rt = (V^2/R)t$
- One kilowatt hour is the energy dissipated by an appliance working at the rate of 1.0 kW for 1.0 hour.
- A fuse is a safety device that cuts off the current to an appliance if the current becomes too large.

Section 5.2 Questions

1 A small torch has a 3.0 V battery connected to a bulb of resistance 15 Ω.
   (a) Calculate:
      (i) the current in the bulb,
      (ii) the power delivered to the bulb.
   (b) The battery supplies a constant current to the bulb for 2.5 hours. Calculate the total energy delivered to the bulb.

2 The capacity of storage batteries is rated in ampere hours (A h). An 80 A h battery can supply a current of 80 A for 1 hour, or 40 A for 2 hours, and so on. Calculate the total energy, in J, stored in a 12 V, 80 A h car battery.

3 An electric kettle is rated at 2.2 kW, 240 V. The supply voltage is reduced from 240 V to 230 V. Calculate the new power of the kettle.

4 In order to save energy and money, the 60 W filament lamps in a room are replaced by 'energy-saving' lamps, each of power 11 W. There are five lamps in the room and, on average, the lamps are in use for 7 hours each day. Electrical energy costs 7.4 pence per kW h. Calculate the money saved during 1 year.

5.3 More about resistance

At the end of Section 5.3 you should be able to:

▥ state and use Ohm's law

▥ determine and describe the *I–V* characteristics of a resistor at constant temperature, a filament lamp and a light-emitting diode (LED)

▥ describe the uses and benefits of LEDs

▥ describe how the resistance of a pure metal and of a negative temperature coefficient (NTC) thermistor is affected by temperature

▥ define *resistivity*

▥ select and use the equation $R = \rho l/A$.

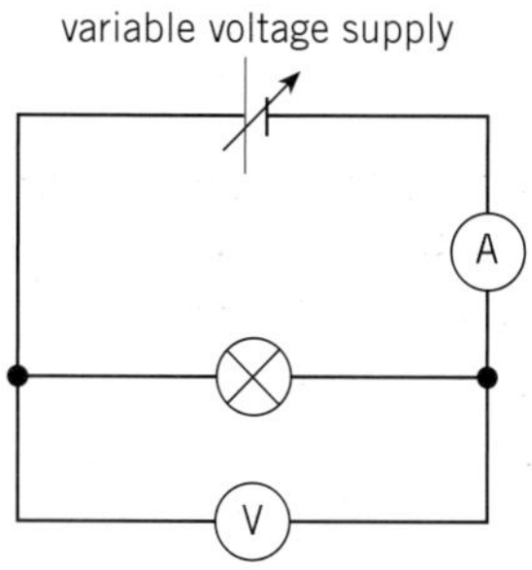

Ohm's law

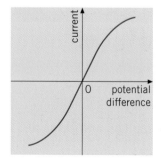

Figure 5.9 Circuit for plotting graphs of current against voltage

Figure 5.10 Current against voltage graph for a filament lamp

The relation between the potential difference across an electrical component and the current through it may be investigated using the circuit of Figure 5.9. For example, if a filament lamp is to be investigated, adjust the voltage across the lamp and measure the corresponding currents and voltages. The variation of current with potential difference is shown in Figure 5.10.

The resistance R of the lamp can be calculated from $R = V/I$. At first the resistance is constant (where the graph is a straight line), but then the resistance increases with current (where the graph curves).

If the lamp is replaced by a length of constantan wire, the graph of the results is as shown in Figure 5.11. It is a straight line through the origin. This shows that for constantan wire, the current is proportional to the voltage. The resistance of the wire is found to stay the same as the current increases. The difference between Figures 5.10 and 5.11 is that the temperature of the constantan wire was constant for all currents used in the experiment, whereas the temperature of the filament of the lamp increased to about 1500 °C.

Graphs like Figure 5.11. would be obtained for wires of any metal, provided that the temperature of the wires did not change during the experiment. The graph illustrates a law discovered by the German scientist Georg Simon Ohm (Figure 5.12). (Ohm's name is now used for the unit of resistance.)

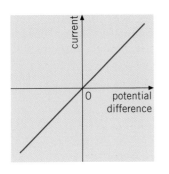

Ohm's law states that, for a conductor at constant temperature, the current in the conductor is proportional to the potential difference across it.

Figure 5.11 Current against voltage graph for a constantan wire

Conductors whose current against voltage graphs are a straight line through the origin, like that in Figure 5.11, are said to obey Ohm's law. It is found that Ohm's law applies to metal wires, provided that the current is not too large. What does 'too large' mean here? It means that the current must not be so great that there is a pronounced heating effect, causing an increase in temperature of the wire.

A lamp filament consists of a thin metal wire. Why does it not obey Ohm's law? (Figure 5.10 shows that the current against voltage graph is not a straight line.) This is because, as stated previously, the temperature of the filament does not remain constant. The increase in current causes the temperature to increase so much that the filament glows.

Figure 5.12 Georg Ohm

Current–voltage graphs

When other devices are tested in the same way, current–voltage graphs like those in Figures 5.13 to 5.15 are obtained.

Thermistors are used as one type of electrical resistance thermometer and are made from semiconducting material. If a thermistor is held at constant temperature in a water bath, its current–voltage graph is a straight line. At a higher temperature, the gradient is steeper, showing that the resistance is lower. Figure 5.13 illustrates this. The resistance of a sample of semiconductor decreases as the temperature increases. This is the opposite behaviour to a metal: the resistance of a metal wire increases as the temperature increases. Thermistors are used as temperature sensors, making use of the fact that their resistance changes considerably with temperature.

Light-dependent resistors (LDRs) are also made from semiconducting material. As their name suggests, their resistance can be altered by a change in light intensity: the brighter the light, the lower the resistance (Figure 5.14). LDRs are used in light sensors and camera exposure meters.

Light-emitting diodes (LEDs) are also made from semiconducting material. The LED conducts when the current is in the direction of the arrowhead on the symbol. This condition is called **forward bias**. When the voltage is reversed, there is negative bias. This is called **reverse bias**. Figure 5.15 shows this important difference in the current–voltage graph when the voltage is reversed. The LED emits light when it conducts, that is, when it is forward biased. A common use of LEDS is as a replacement for filament lamps in indicator panels, particularly for

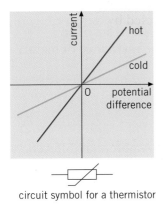

circuit symbol for a thermistor

Figure 5.13 Current against voltage graphs for a thermistor

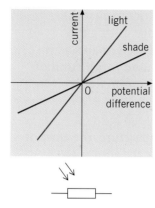

circuit symbol for a light-dependent resistor

Figure 5.14 Current against voltage graphs for a light-dependent resistor

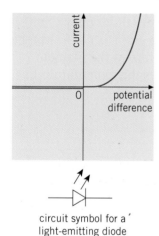

circuit symbol for a light-emitting diode

Figure 5.15 Current against voltage graph for a light-emitting diode

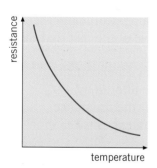

Figure 5.16 Variation with temperature of an NTC thermistor

'standby' situations. LEDs have the following advantages over filament lamps:

▮ robust
▮ small size
▮ reliable
▮ low power dissipation
▮ can be used for very intense light sources.

Resistance and temperature

All solids are made up of atoms that constantly vibrate about their equilibrium positions. The higher the temperature, the greater the amplitude of vibration.

Electric current is the flow of free electrons through the material. As the electrons move, they collide with the vibrating atoms, so their movement is impeded. The more the atoms vibrate, the greater is the chance of collision. This means that the current is less and the resistance is greater.

A temperature rise can cause an increase in the number of free electrons. If there are more electrons free to move, this may outweigh the effect due to the vibrating atoms, and thus the flow of electrons, or the current, will increase. The resistance is therefore reduced. This is the case in semiconductors. Insulators, too, show a reduction in resistance with temperature rise.

Negative temperature coefficient (NTC) thermistors are semiconductor materials whose resistance decreases with temperature rise, as illustrated in Figure 5.16. Typically, at 5 °C the resistance may be 2000 Ω, and at 90 °C only 300 Ω.

For metals there is no increase in the number of free electrons. The increased amplitude of vibration of the atoms makes the resistance of metals increase with temperature.

Resistivity

All materials have some resistance to a flow of charge. A potential difference across the material causes free charges inside to accelerate. As the charges move through the material, they collide with the atoms of the material which get in their way. They transfer some or all of their kinetic energy, and then accelerate again. It is this transfer of energy on collision that causes electrical heating.

As you might guess, the longer a wire, the greater its resistance. This is because the charges have further to go through the material; there is more chance of collisions with the atoms of the material. In fact the resistance is proportional to the length of the wire, or $R \propto l$.

Also, the thicker a wire is, the smaller its resistance will be. This is because there is a bigger area for the charges to travel through, with less chance of collision. In fact the resistance is inversely proportional to the cross-sectional area of the wire, or $R \propto 1/A$. These relations are illustrated in Figures 5.17 and 5.18.

Figure 5.17 The longer the room, the greater the resistance the waiter meets.

Figure 5.18 The wider the room, the easier it is for the waiter to pass through.

Finally, the resistance depends on the type of material. As previously stated, copper is a good conductor, whereas plastics are good insulators. Putting all of this together gives

$$R = \frac{\rho l}{A}$$

where ρ is a constant for a particular material at a particular temperature. ρ is called the **resistivity** of the material at that temperature.

> The resistivity ρ of a material is numerically equal to the resistance between opposite faces of a cube of the material, of unit length and unit cross-sectional area.

So if $\rho = RA/l$ and if the resistance is in ohms, the cross-sectional area in metres squared and the length in metres, then the resistivity is in ohm metres ($\Omega\,\text{m}$). Remember that A is the cross-sectional area through which the current is passing, not the surface area.

We have already seen that the resistance of a wire depends on temperature. Thus, resistivity also depends on temperature. The resistivities of metals increase with increasing temperature, and the resistivities of semiconductors decrease very rapidly with increasing temperature.

The values of the resistivity of some materials at room temperature are given in Table 5.2.

Table 5.2 Resistivities at room temperature

material	resistivity/Ω m
metals:	
copper	1.7×10^{-8}
gold	2.4×10^{-8}
aluminium	2.6×10^{-8}
semiconductors:	
germanium (pure)	0.6
silicon (pure)	2.3×10^3
insulators:	
glass	about 10^{12}
perspex	about 10^{13}
polyethylene	about 10^{14}
sulfur	about 10^{15}

Note the enormous range of resistivities spanned by the materials in this list – a range of 23 orders of magnitude, from $10^{-8}\,\Omega$ m to $10^{15}\,\Omega$ m.

Note, too, that the resistivity is a property of a material, while the resistance is a property of a particular wire or device.

Example

Calculate the resistance per metre at room temperature of a constantan wire of diameter 1.25 mm. The resistivity of constantan at room temperature is $5.0 \times 10^{-7}\,\Omega$ m.

The cross-sectional area of the wire is calculated using πr^2.

Area $= \pi (1.25 \times 10^{-3}/2)^2$

(Don't forget to change the units from mm to m.)

The resistance per metre is given by R/l, and $R/l = \rho/A$. So
resistance per metre $= 5.0 \times 10^{-7}/\pi (1.25 \times 10^{-3}/2)^2$
$$= \mathbf{0.41\ \Omega\ m^{-1}}$$

Now it's your turn

1 Find the length of copper wire, of diameter 0.63 mm, which has a resistance of $1.00\,\Omega$. The resistivity of copper at room temperature is $1.7 \times 10^{-8}\,\Omega$ m.

2 Find the diameter of a copper wire which has the same resistance as an aluminium wire of equal length and diameter 1.20 mm. The resistivities of copper and aluminium at room temperature are $1.7 \times 10^{-8}\,\Omega$ m and $2.6 \times 10^{-8}\,\Omega$ m respectively.

Section 5.3 Summary

- Ohm's law: for a conductor at constant temperature, the current in the conductor is proportional to the potential difference across it.
- The resistance of a metallic conductor increases with increasing temperature; the resistance of a semiconductor decreases with increasing temperature.
- A diode has a low resistance when connected in forward bias, and a very high resistance in reverse bias.
- Resistivity ρ of a material is numerically equal to the resistance between opposite faces of a unit cube of the material: $R = \rho l/A$

Section 5.3 Questions

1 The element of an electric kettle has resistance $26\,\Omega$ at room temperature. The element is made of nichrome wire of diameter 0.60 mm and resistivity $1.1 \times 10^{-6}\,\Omega$ m at room temperature. Calculate the length of the wire.

2 These are values of the current I through an electrical component for different potential differences V across it:

V/V   0   0.19   0.48   1.47   2.92   4.56   6.56   8.70
I/A   0   0.20   0.40   0.60   0.80   1.00   1.20   1.40

(a) Draw a diagram of the circuit that could be used to obtain these values.

(b) Calculate the resistance of the component at each value of current.

(c) Plot a graph to show the variation with current of the resistance of the component.

(d) Suggest what the component is likely to be, giving a reason for your answer.

3 The current in a 2.50 m length of wire of diameter 1.5 mm is 0.65 A when a potential difference of 0.40 V is applied between its ends. Calculate:

(a) the resistance of the wire,

(b) the resistivity of the material of the wire.

5.4 Electrical circuits

At the end of Section 5.4 you should be able to:

- recall and use appropriate circuit symbols
- draw and interpret circuit diagrams
- state Kirchhoff's first law and appreciate that this is a consequence of conservation of charge
- define *electromotive force* (e.m.f.)
- describe the difference between e.m.f. and p.d.
- state Kirchhoff's second law and appreciate that this is a consequence of conservation of energy
- explain what is meant by *internal resistance* and *terminal p.d.*
- select and use the equations $E = I(R + r)$ and $E = V + Ir$.

When reporting an electrical experiment, or describing a circuit, it is essential to know exactly how the components are connected. This could be done by taking a photograph, but this has disadvantages; the photograph in Figure 5.7, for example, is not clear and does not show all the components. Besides, you may not always have a camera with you. You could sketch a block diagram, in which the components are indicated as rectangular boxes labelled 'cell', 'ammeter', 'resistor', etc. The blocks would then be connected with lines to indicate the wiring. This is also unsatisfactory; it takes a lot of time to label all the boxes. It is much better to draw the circuit diagram using a set of symbols recognised by everyone, which do not need to be labelled.

Figure 5.19 shows the symbols that you are likely to need in school and college work, and which you will meet in examination questions. (You will have met many of them already.) It is important that you learn these so that you can recognise them straight away. The only labels you are likely to see on them will be the values of the components, for example 1.5 V for a cell or 22 Ω for a resistor.

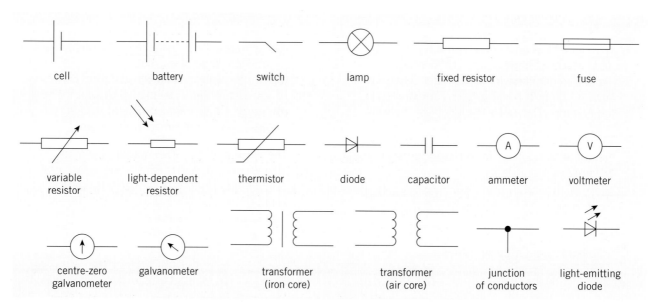

Figure 5.19 Circuit symbols

Series and parallel circuits: Kirchhoff's first law

A **series circuit** is one in which the components are connected one after another, forming one complete loop. You have probably connected an ammeter at different points in a series circuit to show that it reads the same current at each point (Figure 5.20).

A **parallel circuit** is one where the current can take alternative routes in different loops. In a parallel circuit, the current divides at a junction, but the current entering the junction is the same as the current leaving it (Figure 5.21). The fact that the current does not get 'used up' at a junction is because current is the rate of flow of charge, and charges cannot accumulate or get 'used up' at a junction. The consequence of this conservation of electric charge is known as **Kirchhoff's first law**. This law is usually stated as

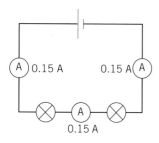

Figure 5.20 The current at each point in a series circuit is the same.

The sum of the currents entering a junction in a circuit is always equal to the sum of the currents leaving it.

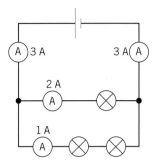

Figure 5.21 The current divides in a parallel circuit.

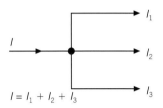

Figure 5.22

At the junction shown in Figure 5.22,

$$I = I_1 + I_2 + I_3$$

Example

For the circuit of Figure 5.23, state the readings of the ammeters A_1, A_2 and A_3.

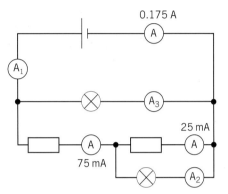

Figure 5.23

A_1 would read **175 mA**, as the current entering the power supply must be the same as the current leaving it.
A_2 would read $75 - 25 = $ **50 mA**, as the total current entering a junction is the same as the total current leaving it.
A_3 would read $175 - 75 = $ **100 mA**.

Now it's your turn

1   The lamps in Figure 5.24 are identical. There is a current of 0.50 A through the battery. What is the current in each lamp?

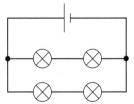

Figure 5.24

2   Figure 5.25 shows one junction in a circuit. Calculate the ammeter reading.

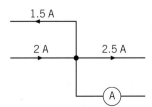

Figure 5.25

Electromotive force and potential difference

When charge passes through a power supply such as a battery, it **gains** electrical energy. The power supply is said to have an **electromotive force**, or **e.m.f.** for short. The electromotive force measures, in volts, the electrical energy gained by each coulomb of charge that passes through the power supply. Note that, in spite of its name, the e.m.f. is *not* a force. The energy gained by the charge comes from the chemical energy of the battery.

$$e.m.f. = \frac{(energy\ converted\ from\ other\ forms\ to\ electrical)}{charge}$$

When charge passes through a resistor, its electrical energy is converted to heat energy in the resistor. The resistor has a **potential difference (p.d.)** across it. The potential difference measures, in volts, the electrical energy lost by each coulomb of charge that passes through it.

$$p.d. = \frac{(energy\ converted\ from\ electrical\ to\ other\ forms)}{charge}$$

Conservation of energy: Kirchhoff's second law

Charge flowing round a circuit gains electrical energy on passing through the battery and loses electrical energy on passing through the rest of the circuit. From the law of conservation of energy, we know that the total energy must remain the same. The consequence of this conservation of energy is known as **Kirchhoff's second law**. This law may be stated as

The sum of the electromotive forces in a closed circuit is equal to the sum of the potential differences.

Figure 5.26 shows a circuit containing a battery, lamp and resistor in series. Applying Kirchhoff's second law, the electromotive force in the circuit is the e.m.f. E of the battery. The sum of the potential differences is the p.d. V_1 across the lamp plus the p.d. V_2 across the resistor. Thus, $E = V_1 + V_2$. If the current in the circuit is I and the resistances of the lamp and resistor are R_1 and R_2 respectively, the p.d.s can be written as $V_1 = IR_1$ and $V_2 = IR_2$, so $E = IR_1 + IR_2$.

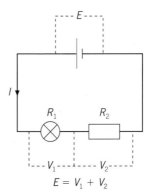

Figure 5.26

$$E = V_1 + V_2$$

It should be remembered that both electromotive force and potential difference have direction. This must be considered when working out the equation for Kirchhoff's second law. For example, in the circuit of Figure 5.27 two cells have been connected in opposition. Here the total electromotive force in the circuit is $E_1 - E_2$, and by Kirchhoff's second law $E_1 - E_2 = V_1 + V_2 = IR_1 + IR_2$.

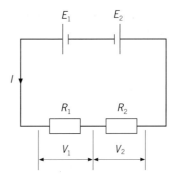

Figure 5.27

Internal resistance

When a car engine is started while the headlights are switched on, the headlights sometimes dim. This is because the car battery has resistance.

All power supplies have some resistance between their terminals, called **internal resistance**. This causes the charge circulating in the circuit to dissipate some electrical energy in the power supply itself. The power supply becomes warm when it delivers a current.

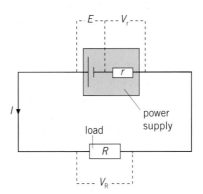

Figure 5.28

Figure 5.28 shows a power supply which has e.m.f. E and internal resistance r. It delivers a current I when connected to an external resistor of resistance R (called the load). V_R is the potential difference across the load, and V_r is the potential difference across the internal resistance. By Kirchhoff's second law,

$$E = V_R + V_r$$

The potential difference V_R across the load is thus given by

$$V_R = E - V_r$$

V_R is called the **terminal potential difference**.

> The terminal potential difference is the p.d. between the terminals of a cell when a current is being delivered.

The terminal potential difference is always less than the electromotive force when the power supply delivers a current. This is because of the potential difference across the internal resistance. The potential difference across the internal resistance is sometimes called the **lost volts**.

> *lost volts = e.m.f. – terminal p.d.*

In contrast, the electromotive force is the terminal potential difference when the cell is on open circuit (when it is delivering no current). This e.m.f. may be measured by connecting a very high resistance voltmeter across the terminals of the cell.

You can use the circuit in Figure 5.29 to show that the greater the current delivered by the power supply, the lower its terminal potential difference. As more lamps are connected in parallel to the power supply, the current increases and the lost volts, given by

> *lost volts = current × internal resistance*

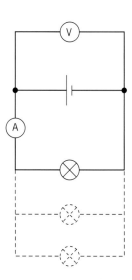

Figure 5.29 Effect of circuit current on terminal potential difference

will increase. Thus the terminal potential difference decreases.

To return to the example of starting a car with its headlights switched on, a large current (perhaps 100 A) is supplied to the starter motor by the battery. There will then be a large potential difference across the internal resistance; that is, the lost voltage will be large. The terminal potential difference will drop and the lights will dim.

In the terminology of Figure 5.28, $V_R = IR$ and $V_r = Ir$, so $E = V_R + V_r$ becomes $E = IR + Ir$, or $E = I(R + r)$.

The maximum current that a power supply can deliver will be when its terminals are short-circuited by a wire of negligible resistance, so that $R = 0$. In this case, the potential difference across the internal resistance will equal the e.m.f. of the cell. The terminal p.d is then zero. **Warning: do not try out this experiment, as the wire gets very hot; there is also a danger of the battery exploding.**

Quite often, in problems, the internal resistance of a supply is assumed to be negligible, so that the potential difference V_R across the load is equal to the e.m.f. of the power supply.

Effect of internal resistance on power from a battery

The power delivered by a battery to a variable external load resistance can be investigated using the circuit of Figure 5.30. Readings of current I and potential difference V_R across the load are taken for different values of the variable load resistor. The product $V_R I$ gives the power dissipated in the load, and the quotient V_R/I gives the load resistance R.

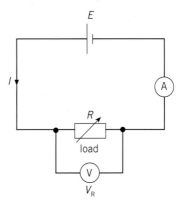

Figure 5.30 Circuit for investigating power transfer to an external load

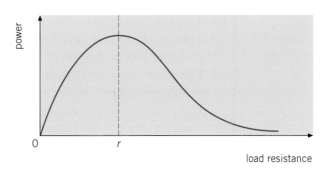

Figure 5.31 Graph of power delivered to external load against load resistance

Figure 5.31 shows the variation with load resistance R of the power $V_R I$ dissipated. The graph indicates that there is a maximum power delivered by the battery at one value of the external resistance. This value is equal to the internal resistance r of the battery.

A battery delivers maximum power to a circuit when the load resistance of the circuit is equal to the internal resistance of the battery.

Example

A high-resistance voltmeter reads 13.0 V when it is connected across the terminals of a battery. The voltmeter reading drops to 12.0 V when the battery delivers a current of 3.0 A to a lamp. State the e.m.f. of the battery. Calculate the potential difference across the internal resistance (the lost volts) when the battery is connected to the lamp. Calculate the internal resistance of the battery.

The e.m.f. is **13.0 V**, since this is the voltmeter reading when the battery is delivering negligible current.

Using $V_r = E - V_R$, lost volts = V_r = 13.0 – 12.0 = **1.0 V**.

Using $V_r = Ir$, r = 1.0/3.0 = **0.33 Ω**.

Now it's your turn

1   Three cells, each of e.m.f. 1.5 V, are connected in series to a 15 Ω light bulb. The current in the circuit is 0.27 A. Calculate the internal resistance of each cell.

2   A cell of e.m.f. 1.5 V has an internal resistance of 0.50 Ω. Calculate the maximum current it can deliver. Under what circumstances does it deliver this maximum current? Calculate also the maximum power it can deliver to an external load. Under what circumstances does it deliver this maximum power?

Section 5.4 Summary

▦   At any junction in a circuit, the total current entering the junction is equal to the current leaving it. This is Kirchhoff's first law, and is a consequence of the law of conservation of charge.

▦   The electromotive force (e.m.f.) of a supply measures the electrical energy gained per unit of charge passing through the supply.

▦   The potential difference (p.d.) across a resistor measures the electrical energy converted per unit of charge passing through the resistor.

▦   In any closed loop of a circuit, the sum of the electromotive forces is equal to the sum of the potential differences. This is Kirchhoff's second law, and is a consequence of the law of conservation of energy.

▦   The voltage across the terminals of a supply (the terminal p.d.) is always less than the e.m.f. of the supply when the supply is delivering a current, because of the lost volts across the internal resistance.

▦   For a supply of e.m.f. E which has internal resistance r, $E = I(R + r)$ where R is the external circuit resistance and I is the current in the supply.

▦   A supply delivers maximum power to a load when the load resistance is equal to the internal resistance of the supply.

Section 5.4 Questions

1 The internal resistance of a dry cell increases gradually with age, even if the cell is not being used. However, the e.m.f. remains approximately constant. You can check the age of a cell by connecting a low-resistance ammeter across the cell and measuring the current. For a new 1.5 V cell of a certain type, the short-circuit current should be about 30 A.
   (a) Calculate the internal resistance of a new cell.
   (b) A student carries out this test on an older cell, and finds the short-circuit current to be only 5 A. Calculate the internal resistance of this cell.

2 A torch bulb has a power supply of two 1.5 V cells connected in series. The potential difference across the bulb is 2.2 V, and it dissipates energy at the rate of 550 mW. Calculate:
   (a) the current through the bulb,
   (b) the internal resistance of each cell,
   (c) the heat energy dissipated in each cell in two minutes.

3 Two identical light bulbs are connected first in series, and then in parallel, across the same battery (assumed to have negligible internal resistance). Use Kirchhoff's laws to decide which of these connections will give the greater total light output.

5.5 More on series and parallel circuits

At the end of Section 5.5 you should be able to:
- select and use the equation for the total resistance of two or more resistors in series
- recall and use the equation for the total resistance of two or more resistors in parallel
- explain how a potential divider circuit can be used to produce a variable p.d.
- recall and use the potential divider equation $V_{out} = V_{in} \times R_2/(R_1 + R_2)$
- describe the use of thermistors and light-dependent resistors in potential divider circuits
- describe the advantages of using data-loggers to monitor physical changes.

Resistors in series

Figure 5.32 shows two resistors of resistances R_1 and R_2 connected in series, and a single resistor of resistance R equivalent to them. The current I in the resistors, and in their equivalent single resistor, is the same.

The total potential difference V across the two resistors must be the same as that across the single resistor. If V_1 and V_2 are the potential differences across each resistor,

$$V = V_1 + V_2$$

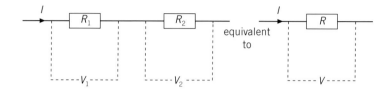

Figure 5.32 Resistors in series

But since potential difference is given by multiplying the current by the resistance,

$$IR = IR_1 + IR_2$$

Dividing by the current I,

$$R = R_1 + R_2$$

This equation can be extended so that the equivalent resistance R of several resistors connected in series is given by the expression

$$R = R_1 + R_2 + R_3 + \cdots$$

Thus

The combined resistance of resistors in series is the sum of all the individual resistances.

Resistors in parallel

Now consider two resistors of resistance R_1 and R_2 connected in parallel, as shown in Figure 5.33. The current through each will be different, but they will each have the same potential difference. The equivalent single resistor of resistance R will have the same potential difference across it, but the current will be the total current through the separate resistors.

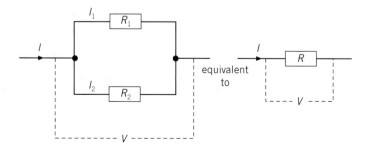

Figure 5.33 Resistors in parallel

By Kirchhoff's first law,

$I = I_1 + I_2$

and, using Ohm's law, $I = V/R$, so

$V/R = V/R_1 + V/R_2$

Dividing by the potential difference V,

$1/R = 1/R_1 + 1/R_2$

This equation can be extended so that the equivalent resistance R of several resistors connected in parallel is given by

$$\frac{1}{R} = \frac{1}{R_1} + \frac{1}{R_2} + \frac{1}{R_3} + \cdots$$

Thus

> The reciprocal of the combined resistance of resistors in parallel is the sum of the reciprocals of all the individual resistances.

Note that:

1 For two identical resistors in parallel, the combined resistance is equal to half of the value of each one.
2 For resistors in parallel, the combined resistance is always less than the value of the smallest individual resistance.

Example

Calculate the equivalent resistance of the arrangement of resistors in Figure 5.34.

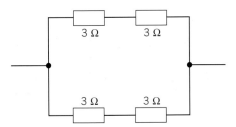

Figure 5.34

The arrangement is equivalent to two $6\,\Omega$ resistors in parallel, so the combined resistance R is given by $1/R = 1/6 + 1/6 = 2/6$. (Don't forget to find the reciprocal of this value.)

Thus $R = \mathbf{3\,\Omega}$.

Now it's your turn

1 Calculate the equivalent resistance of the arrangement of resistors in Figure 5.35. *Hint*: First find the resistance of the parallel combination.

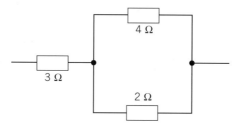

Figure 5.35

2 Calculate the effective resistance between the points A and B in the network in Figure 5.36.

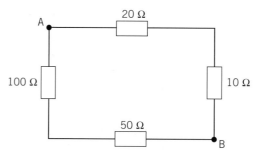

Figure 5.36

Potential dividers and potentiometers

Two resistors connected in series with a cell each have a potential difference. They may be used to divide the e.m.f. of the cell. This is illustrated in Figure 5.37.

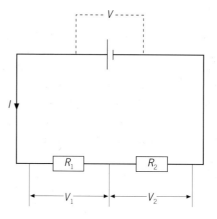

Figure 5.37 The potential divider

The current in each resistor is the same, because they are in series. Thus $V_1 = IR_1$ and $V_2 = IR_2$. Dividing the first equation by the second gives $V_1/V_2 = R_1/R_2$. The ratio of the voltages across the two resistors is the same as the ratio of their resistances. If the potential difference across the combination were 12 V and R_1 were equal to R_2, then each resistor would have 6 V across it. If R_1 were twice the magnitude of R_2, then V_1 would be 8 V and V_2 would be 4 V.

A **potentiometer** is a continuously variable potential divider. In Section 5.3, a variable voltage supply was used to vary the voltage across different circuit components. A variable resistor, or rheostat, may be used to produce a continuously variable voltage.

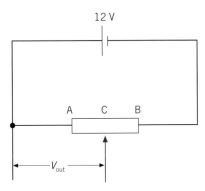

Figure 5.38 Potentiometer circuit

Such a variable resistor is shown in Figure 5.38. The fixed ends AB are connected across the battery so that there is the full battery voltage across the whole resistor. As with the potential divider, the ratio of the voltages across AC and CB will be the same as the ratio of the resistances of AC and CB. When the sliding contact C is at the end B, the output voltage V_{out} will be 12 V. When the sliding contact is at end A, then the output voltage will be zero. So, as the sliding contact is moved from A to B, the output voltage varies continuously from zero up to the battery voltage. In terms of the terminal p.d. V of the cell, the output V_{out} of the potential divider is given by

$$V_{out} = \frac{VR_1}{(R_1 + R_2)}$$

A variable resistor connected in this way is called a potentiometer. A type of potentiometer is shown in Figure 5.39. Note the three connections.

If a device with a variable resistance is connected in series with a fixed resistor, and the combination is connected to a cell or battery to make a potential divider, then we have the situation of a potential divider that is variable between certain limits. The device of variable resistance could be, for example, a light-dependent resistor or a thermistor, as shown in Section 5.3. Changes in the illumination or the temperature cause a change in the resistance of one component of the potential divider, so that the potential difference across this component changes. The change in the potential difference can be used to operate control circuitry if, for example, the illumination becomes too low or too high, or the temperature falls outside certain limits.

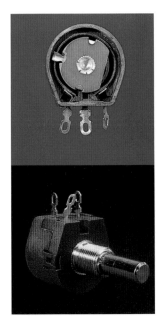

Figure 5.39 Internal and external views of a potentiometer

A potentiometer can also be used as a means of comparing potential differences. The circuit of Figure 5.40 illustrates the principle. In this circuit the variable potentiometer resistor consists of a length of uniform resistance wire, stretched along a metre rule. Contact can be made to any point on this wire using a sliding contact. Suppose that the cell A has a known e.m.f. E_A. This cell is switched into the circuit using the two-way switch. The sliding contact is then moved along the wire until the centre-zero galvanometer reads zero. The length l_A of the wire from the common zero end to the sliding contact is noted. Cell B has an unknown e.m.f. E_B. This cell is then switched into the circuit and the balancing process repeated. Suppose that the position at which the galvanometer reads zero is then a distance l_B from the common zero to the sliding contact. The ratio of the e.m.f.s is the ratio of the balance lengths; that is, $E_B/E_A = l_B/l_A$, and E_B can be determined in terms of the known e.m.f. E_A.

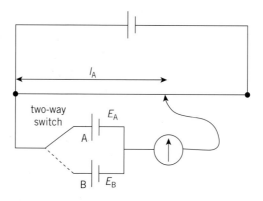

Figure 5.40 Potentiometer used to compare cell e.m.f.s

The use of data-loggers

The potential divider provides a means by which a change in an environmental condition may be converted into a change in potential difference. For example, use of a thermistor enables temperatures to be monitored as potential differences.

A data-logger, used in conjunction with a computer, enables the monitoring of quantities that vary with time by storing the voltage measurements. The quantities that can be monitored include temperature, light intensity, pH, humidity, position and stress.

The data-logger is connected between the sensor and a computer. When switched on, the data-logger monitors and stores the data as a function of time. The computer provides a read-out showing the variation with time of the quantity being monitored.

Data-loggers provide a versatile means of measurement. Their advantages include:

▦ they can measure the variation with time of the quantity
▦ the variation can be over very short times, for example the position of a pendulum bob, or over very long times, for example the stretching over a period of a day of an elastic band under constant load
▦ they can be remote sensing

there is no need for an observer to be present

they enable processing of data, for example data collected for variation with time of position can be processed to give variation with time of speed or acceleration.

Example

1 A light-emitting diode (LED) is connected in series with a resistor to a 5.0 V supply.
   (a) Calculate the resistance of the series resistor required to give a current in the LED of 12 mA, with a voltage across it of 2.0 V.
   (b) Calculate the potential difference across the LED when the series resistor has resistance 500 Ω. The resistance of the LED is the same as that in **(a)**.

   (a) If the supply voltage is 5.0 V and the p.d. across the LED is 2.0 V, the p.d. across the resistor must be 5.0 – 2.0 = 3.0 V. The current through the resistor is 12 mA as it is in series with the LED. Using $R = V/I$, the resistance of the resistor is $3.0/12 \times 10^{-3} = $ **250 Ω**.
   (b) The resistance of the LED is given by $R = V/I = 2.0/12 \times 10^{-3} = 167\,\Omega$. If this resistance is in series with a 500 Ω resistor and a 5.0 V supply, the p.d. across the LED is $5.0 \times 167/(167 + 500) = $ **1.25 V**.

2 The e.m.f.s of two cells are compared using the slide-wire circuit of Figure 5.40. Cell A has a known e.m.f. of 1.02 V; using this cell, a balance point is obtained when the slider is 37.6 cm from the zero of the scale. Using cell B, the balance point is at 55.3 cm.
   (a) Calculate the e.m.f. of cell B.
   (b) State the advantage of using this null method to compare the e.m.f.s.

   (a) This is a straightforward application of the formula for the potentiometer, $E_B/E_A = l_B/l_A$. Substituting the values, $E_B = $ **1.50 V**.
   (b) When comparing the e.m.f.s of cells, it is necessary to arrange for the cells to be on open circuit so that there is no drop in terminal potential difference because of a current passing through the internal resistance. When the potentiometer is balanced, there is no current from the cell under test, which is exactly what is required.

Now it's your turn

1 Figure 5.41 shows a light-dependent resistor (LDR) connected in series with a 10 kΩ resistor and a 12 V supply. Calculate:
   (a) the p.d. V_L across the LDR when it is in the dark and has resistance 8.0 MΩ,

(b) the p.d. V_L across the LDR when it is in bright light and has resistance $500\,\Omega$,

(c) the resistance of the LDR in lighting conditions which make V_L equal to $4.0\,V$.

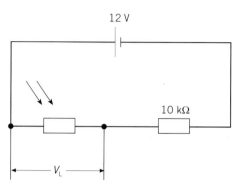

Figure 5.41

2 The thermistor in the potential divider circuit of Figure 5.42 has a resistance which varies between $100\,\Omega$ at $100\,°C$ and $6.0\,k\Omega$ at $0\,°C$. Calculate the potential difference across the thermistor at

(a) $100\,°C$, **(b)** $0\,°C$.

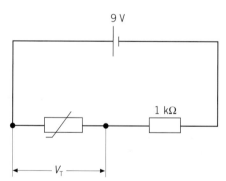

Figure 5.42

Section 5.5 Summary

▦ The equivalent resistance R of resistors connected in series is given by: $R = R_1 + R_2 + R_3 + \cdots$

▦ The equivalent resistance R of resistors connected in parallel is given by: $1/R = 1/R_1 + 1/R_2 + 1/R_3 + \cdots$

▦ Two resistors in series act as a potential divider, where $V_1/V_2 = R_1/R_2$. If V is the supply voltage: $V_{out} = VR_1/(R_1 + R_2)$

▦ A potentiometer is a variable resistor connected as a potential divider to give a continuously variable output voltage.

Section 5.5 Questions

1 You are given three resistors of resistance 22 Ω, 47 Ω and 100 Ω. Calculate:
(a) the maximum possible resistance,
(b) the minimum possible resistance,
that can be obtained by combining any or all of these resistors.

2 In the circuit of Figure 5.43, the currents I_1 and I_2 are equal. Calculate:
(a) the resistance R of the unknown resistor,
(b) the total current I_3.

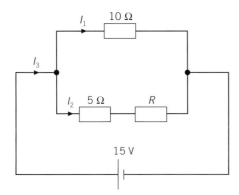

Figure 5.43

3 Figure 5.44 shows a potential divider circuit, designed to provide p.d.s of 1.0 V and 4.0 V from a battery of e.m.f. 9.0 V and negligible internal resistance.
(a) Calculate the value of resistance R.
(b) State and explain what happens to the voltage at terminal A when an additional 1.0 Ω resistor is connected between terminals B and C in parallel with the 5.0 Ω resistor. No calculations are required.

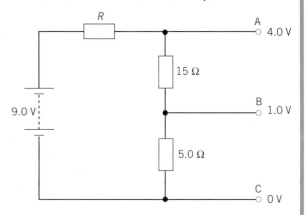

Figure 5.44

Exam-style Questions

1 A student designs an electrical method to monitor the position of a steel sphere rolling on two parallel rails. Each rail is made from bare wire of length 30 cm and resistance 20 Ω. The position-sensing circuit is shown in Figure 5.45. The resistance of the steel sphere and the internal resistance of the battery are negligible.

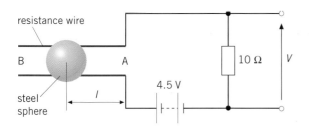

Figure 5.45

(a) State the voltage across the 10 Ω resistor when the sphere is at A, where $I = 0$.
(b) With the sphere at end B of the rails, calculate:
  (i) the total resistance of the circuit,
  (ii) the current in the 10 Ω resistor,
  (iii) the output voltage V.

2 Two equations for the power P dissipated in a resistor are $P = I^2R$ and $P = V^2/R$. The first suggests that the greater the resistance R of the resistor, the more power is dissipated. The second suggests the opposite: the greater the resistance, the less the power. Explain this inconsistency.

3 State the minimum number of resistors, each of the same resistance and power rating of 0.5 W, which must be used to produce an equivalent 1.2 kΩ, 5 W resistor. Calculate the resistance of each, and state how they should be connected.

4 In the circuit shown in Figure 5.46 the current in the battery is 1.5 A. The battery has internal resistance 1.0 Ω. Calculate:
   (a) the combined resistance of the resistors that are connected in parallel in the circuit of Figure 5.46,
   (b) the total resistance of the circuit,
   (c) the resistance of resistor Y,
   (d) the current through the 6 Ω resistor.

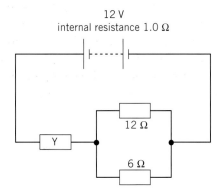

12 V
internal resistance 1.0 Ω

Y

12 Ω

6 Ω

Figure 5.46

5 The current in the starter motor of a car is 160 A when starting the engine. The connecting cable has total length 1.3 m, and consists of fifteen strands of wire, each of diameter 1.2 mm. The resistivity of the metal of the strands is $1.4 \times 10^{-8}\,\Omega\,m$.
   (a) Calculate:
      (i) the resistance of each strand,
      (ii) the total resistance of the cable,
      (iii) the power loss in the cable.

(b) When the starter motor is used to start the car, 700 C of charge pass through a given cross-section of the cable.
   (i) Assuming that the current is constant at 160 A, calculate for how long the charge flows.
   (ii) Calculate the number of electrons which pass a given cross-section of the cable in this time. The electron charge e is $-1.6 \times 10^{-19}\,C$.
(c) The e.m.f. of the battery is 13.6 V and its internal resistance is 0.012 Ω. Calculate:
   (i) the potential difference across the battery terminals when the current in the battery is 160 A,
   (ii) the rate of production of heat energy in the battery.

6 A copper wire of length 16 m has a resistance of 0.85 Ω. The wire is connected across the terminals of a battery of e.m.f. 1.5 V and internal resistance 0.40 Ω.
   (a) Calculate the potential difference across the wire and the power dissipated in it.
   (b) In an experiment, the length of this wire connected across the terminals of the battery is gradually reduced.
      (i) Sketch a graph to show how the power dissipated in the wire varies with the connected length.
      (ii) Calculate the length of the wire when the power dissipated in the wire is a maximum.
      (iii) Calculate the maximum power dissipated in the wire.

6. Waves

The aim of this chapter is to introduce some general properties of waves. We shall meet two broad classifications of waves, transverse and longitudinal, based on the direction in which the particles vibrate relative to the direction in which the wave transmits energy. We shall define terms such as amplitude, wavelength and frequency for a wave, and derive the relationship between speed, frequency and wavelength. We shall look at demonstrations of some properties of waves, such as reflection, refraction, diffraction and interference. In connection with interference, we shall use the principle of superposition, which tells us how waves which come together at the same place interact.

6.1 Wave motion

At the end of Section 6.1 you should be able to:
- describe and distinguish between progressive longitudinal and transverse waves
- define and use the terms *displacement*, *amplitude*, *wavelength*, *period*, *phase difference*, *frequency* and *speed* of a wave
- use the relationships *intensity = power / cross-sectional area* and *intensity* $\propto$ *amplitude*^{2}
- derive, select and use the wave equation $v = f\lambda$
- explain what is meant by *reflection*, *refraction* and *diffraction* of waves
- state typical values for the wavelengths of the different regions of the electromagnetic spectrum
- state that all electromagnetic waves travel at the same speed in a vacuum
- describe differences and similarities between regions of the electromagnetic spectrum
- describe some of the practical uses of electromagnetic waves
- describe the characteristics and dangers of UV radiation and explain the role of sunscreen.

Wave motion is a means of moving energy from place to place. For example, electromagnetic waves from the Sun carry the energy that plants need to survive and grow. The energy carried by sound waves causes our ear drums to vibrate. The energy carried by seismic waves (earthquakes) can devastate vast areas, causing land to move and buildings to collapse. Waves which move energy from place to place are called **progressive waves**.

Vibrating objects act as sources of waves. For example, a vibrating tuning fork sets the air close to it into oscillation, and a sound wave spreads out from the fork. For a radio wave, the vibrating objects are electrons.

There are two main groups of waves. These are **transverse waves** and **longitudinal waves**.

A transverse wave is one in which the vibrations of the particles in the wave are at right angles to the direction in which the energy of the wave is travelling.

Figure 6.1 shows a transverse wave moving along a rope. The particles of the rope vibrate up and down, whilst the energy travels at right angles to this, from A to B. There is no transfer of matter from A to B. Examples of transverse waves include light waves, surface water waves and secondary seismic waves (S-waves).

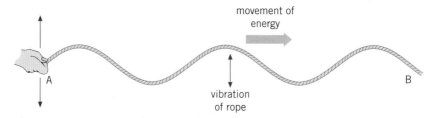

Figure 6.1 Transverse wave on a rope

A longitudinal wave is one in which the direction of the vibrations of the particles in the wave is along the direction in which the energy of the wave is travelling.

Figure 6.2 shows a longitudinal wave moving along a stretched spring (a 'slinky'). The coils of the spring vibrate along the length of the spring, whilst the energy travels along the same line, from A to B. Note that the spring itself does not move from A to B. Examples of longitudinal waves include sound waves and primary seismic waves (P-waves).

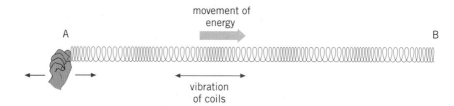

Figure 6.2 Longitudinal wave on a slinky spring

Graphical representation of waves

The **displacement** of a particle on a wave is its distance from its rest position.

Displacement is a vector quantity; it can be positive or negative. A transverse wave may be represented by plotting displacement y on the y-axis against distance x along the wave, in the direction of energy travel, on the x-axis. This is shown in Figure 6.3. It can be seen that the graph is a snapshot of what is actually observed to be a transverse wave.

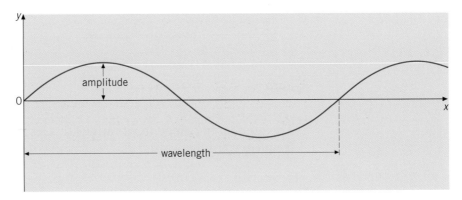

Figure 6.3 Displacement–distance graph for a transverse wave

For a longitudinal wave, the displacement of the particles is along the direction of energy travel. However, if these displacements are plotted on the y-axis of a graph of displacement against distance, the graph has exactly the same shape as for a transverse wave (Figure 6.3). This is very useful, in that one graph can represent both types of wave. Using this graph, wave properties may be treated without reference to the type of wave.

> The **amplitude** of the wave motion is defined as the maximum displacement of a particle in the wave.

Also, it can be seen that the wave repeats itself. That is, the wave can be constructed by repeating a section of the wave. The length of the smallest repetition unit is called the **wavelength**.

> One wavelength is the distance between two neighbouring peaks or two neighbouring troughs, or two neighbouring points which are vibrating together in exactly the same way (in phase). It is the distance moved by the wave during one oscillation of the source of the waves.

The usual symbol for wavelength is λ, the Greek letter lambda.

Another way to represent both waves is to plot a graph of displacement y against time t. This is shown in Figure 6.4. Again, the wave repeats itself after a certain interval of time. The time for one complete vibration is called the **period** of the wave T.

> The period of the wave is the time for a particle in the wave to complete one vibration, or one cycle.

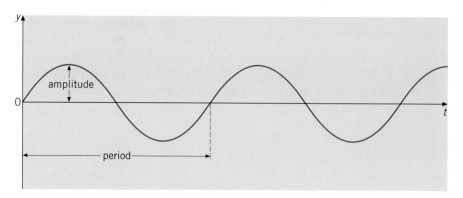

Figure 6.4 Displacement–time graph for a wave

The number of complete vibrations (cycles) per unit time is called the **frequency** f of the wave.

For waves on ropes and springs, displacement and amplitude are measured in mm, m or other units of length. Period is measured in seconds (s). Frequency has the unit per second (s^{-1}) or hertz (Hz).

Application: measurement of frequency using a cathode-ray oscilloscope

A cathode-ray oscilloscope (c.r.o.) has a calibrated time-base, so that measurements from the screen of the c.r.o. can be used to give values of time intervals. One application is to measure the frequency of a periodic signal, for example the sine-wave output of a signal generator or of a microphone. The electrical signal from the generator or from the microphone is connected to the Y-input of the c.r.o., and the Y-amplifier and time-base controls are adjusted until a trace of at least one, but fewer than about five, complete cycles of the signal is obtained on the screen. The distance L on the graticule (the scale on the screen) corresponding to one complete cycle is measured (Figure 6.5). It is good practice to measure the length of, say, four cycles, and then divide by four so as to obtain an average value of L. The graticule will probably be divided into centimetre and perhaps millimetre or

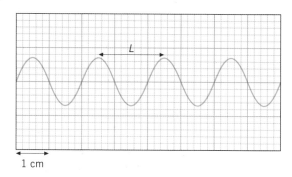

Figure 6.5 Measurement of frequency using a c.r.o.

two-millimetre divisions. If the time-base setting is x (which will be in units of seconds, milliseconds or microseconds per centimetre), the time T for one cycle is given by $T = Lx$. The frequency f of the signal is then obtained from $f = 1/T$. The uncertainty of the determination will depend on how well you can estimate the measurement of the length of the cycle from the graticule. Remembering that the trace has a finite width, you can probably measure this length to an uncertainty of about ±2 mm.

Example

The output of a signal generator is connected to the Y-input of a c.r.o. When the time-base control is set at 0.50 milliseconds per centimetre, the trace shown in Figure 6.6 is obtained. What is the frequency of the signal?

Two complete cycles of the trace occupy 6.0 cm on the graticule. The length of one cycle is therefore 3.0 cm. The time-base setting is 0.50 ms cm^{-1}, so 3.0 cm is equivalent to $3.0 \times 0.50 = 1.5$ ms. The frequency is thus $1/1.5 \times 10^{-3} =$ **670 Hz**.

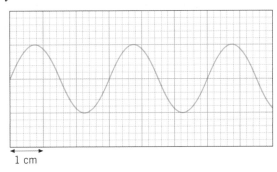

1 cm

Figure 6.6

Now it's your turn

The same signal is applied to the Y-input of the c.r.o. as in the example above, but the time-base control is changed to 2.0 milliseconds per centimetre. How many complete cycles of the trace will appear on the screen, which is 8.0 cm wide?

A term used to describe the relative positions of the crests or troughs of two waves of the same frequency is **phase**. When the crests and troughs of the two waves are aligned, the waves are said to be **in phase**. When a crest is aligned with a trough, the waves are **out of phase**. When used as a quantitative measure, phase has the unit of angle (radians or degrees). Thus, when waves are out of phase, one wave is half a cycle behind the other. Since one cycle is equivalent to 2π radians or 360°, the **phase difference** between waves that are exactly out of phase is π radians or 180°.

Consider Figure 6.7, in which there are two waves of the same frequency, but with a phase difference between them. The period T corresponds to a phase angle of 2π rad or 360°. The two waves are out of step by a time t. Thus, phase difference is equal to $2\pi(t/T)$ rad = $360(t/T)°$. A similar argument may be used for waves of wavelength λ which are out of step by a distance x. In this case the phase difference is $2\pi(x/\lambda)$ rad = $360(x/\lambda)°$.

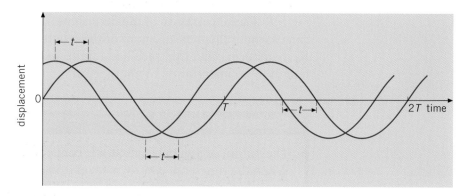

Figure 6.7 Phase difference

One of the characteristics of a progressive wave is that it carries energy. The amount of energy passing through unit area per unit time is called the **intensity** of the wave. Since energy per unit time is power, intensity is power per unit area and is measured in $J\,s^{-1}\,m^{-2}$ or $W\,m^{-2}$. The intensity is proportional to the square of the amplitude of a wave. Thus, doubling the amplitude of a wave increases the intensity of the wave by a factor of four. The intensity also depends on the frequency: intensity is proportional to the square of the frequency.

For a wave of amplitude A and frequency f, the intensity I is proportional to A^2f^2.

If the waves from a point source spread out equally in all directions, we have what is called a **spherical wave**. As the wave travels further from the source, the energy it carries passes through an increasingly large area. Since the surface area of a sphere is $4\pi r^2$, the intensity is $P/4\pi r^2$, where P is the power of the source. The intensity of the wave thus decreases with increasing distance from the source. The intensity I is proportional to $1/r^2$, where r is the distance from the source.

This relationship assumes that there is no absorption of wave energy.

Wave equation

From the definition of wavelength λ, in one cycle of the source the wave energy moves a distance λ. In f cycles, the wave moves a distance $f\lambda$. If f is the frequency of the wave, then f cycles are produced in unit time. Therefore $f\lambda$ is the distance moved in unit time. Referring to Chapter 2, speed v is the distance moved per unit time. Therefore

$$v = f\lambda$$

or

speed = frequency × wavelength

This is an important relationship between the speed of a wave and its frequency and wavelength.

The speed of a progressive wave is the speed at which energy is transferred along the wave.

Examples

1 A tuning fork of frequency 170 Hz produces sound waves of wavelength 2.0 m. Calculate the speed of sound.

Using $v = f\lambda$, we have $v = 170 \times 2.0 =$ **340 m s^{-1}**.

2 The amplitude of a wave in a rope is 15 mm. If the amplitude were changed to 20 mm, keeping the frequency the same, by what factor would the power carried by the rope change?

Intensity is proportional to the square of the amplitude. Here the amplitude has been increased by a factor of 20/15, so the power carried by the wave increases by a factor of $(20/15)^2 =$ **1.8**.

Now it's your turn

1 Water waves of wavelength 0.080 m have a frequency 5.0 Hz. Calculate the speed of these water waves.

2 The speed of sound is 340 m s^{-1}. Calculate the wavelength of the sound wave produced by a violin when a note of frequency 500 Hz is played.

3 The speed of light is 3.0×10^8 m s^{-1}. Calculate the frequency of red light of wavelength 6.5×10^{-7} m.

4 A beam of red light has twice the intensity of another beam of the same colour. Calculate the ratio of the amplitudes of the waves.

Properties of wave motions

Although there are many different types of waves (light waves, sound waves, electrical waves, mechanical waves, etc.) there are some basic properties which they all have in common. All waves can be reflected, refracted, diffracted, and can produce interference patterns.

These properties may be demonstrated using a ripple tank similar to that shown in Figure 6.8. As the motor turns, the wooden bar vibrates, creating ripples on the surface of the water. The ripples are lit from above. This creates shadows on the viewing screen. The shadows show the shape and movement of the waves. Each shadow corresponds to a particular point on the wave, and is referred to as a **wavefront**.

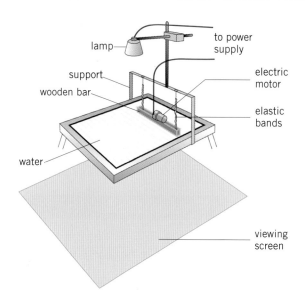

Figure 6.8 Ripple tank

Figure 6.9 illustrates the pattern of wavefronts produced by a low-frequency vibrator and one of higher frequency. Note that for the higher frequency the wavelength is less, since wave speed is constant and $v = f\lambda$.

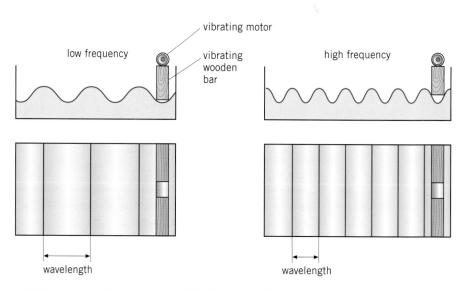

Figure 6.9 Ripple tank patterns for low- and high-frequency vibrations

Circular waves may be produced by replacing the vibrating bar with a small dipper, or by allowing drops of water to fall into the ripple tank. A circular wave is illustrated in Figure 6.10. This pattern is characteristic of waves spreading out from a point source.

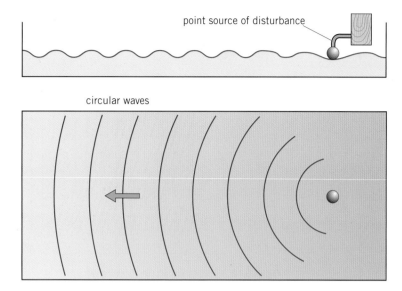

Figure 6.10 Ripple tank pattern for a point source

We shall now see how the ripple tank may be used to demonstrate the wave properties of reflection, refraction, diffraction and interference.

Reflection

As the waves strike a plane barrier placed in the water, they are reflected. The angle of reflection equals the angle of incidence, and there is no change in wavelength (Figure 6.11a). If a curved barrier is used, the waves can be made to converge or diverge (Figure 6.11b).

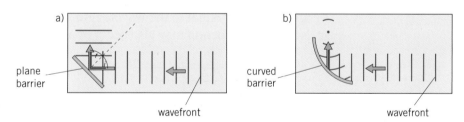

Figure 6.11 Ripple tank pattern showing reflection at a) a plane surface and b) a curved one

The reflection of a sound wave may be detected as an echo. An echo is a sound that is heard for a second time after the sound wave has been reflected from an object (such as a building). You use the reflection of light waves every time you look in a mirror!

Refraction

If a glass block is submerged in the water, this produces a sudden change in the depth of the water. The speed of surface ripples on water depends on the depth of the water: the shallower the water, the slower the speed. Thus, the waves move more slowly as they pass over the glass block. The frequency of the waves remains constant, and so the wavelength decreases. If the waves are incident at an angle to the submerged block, they will change direction, as shown in Figure 6.12.

The change in direction of a wave due to a change in speed is called **refraction**.

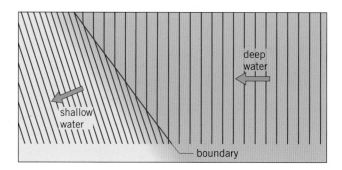

Figure 6.12 Ripple tank pattern showing refraction

As the waves re-enter the deeper water, their speed increases to the former value and they change direction once again.

Refraction of light at the surface of water is what makes the water appear to be shallower than it is in reality. The refraction of sound waves in different densities of sea water causes problems when interpreting the images produced by sonar.

Diffraction

When waves pass through a narrow gap, they spread out. This spreading out is called **diffraction**. The extent of diffraction depends on the width of the gap compared with the wavelength. It is most noticeable if the width of the gap is approximately equal to the wavelength. Diffraction is illustrated in Figure 6.13. Note that diffraction may also occur at an edge.

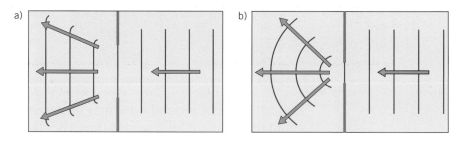

Figure 6.13 Ripple tank pattern showing diffraction at a) a wide gap, b) a narrow gap

Diffraction of sound waves enables you to hear objects that are round a corner although you cannot see them. The wavelength of light is much less than the wavelength of sound and so there is very little diffraction of the light. Diffraction of light will be considered further in Section 6.4.

Diffraction is defined as the spreading of a wave into regions where it would not be seen if it moved only in straight lines.

Electromagnetic waves

Visible light is just a small region of the **electromagnetic spectrum**. All electromagnetic waves are transverse waves, consisting of electric and magnetic fields which oscillate at right angles to each other and to the direction in which the wave is travelling. This is illustrated in Figure 6.14.

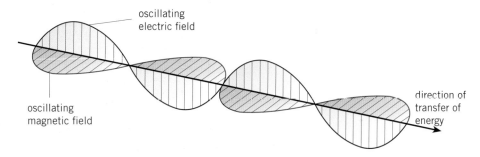

Figure 6.14 Oscillating electric and magnetic fields in an electromagnetic wave

Electromagnetic (e.m.) waves show all the properties common to wave motions: they can be reflected, refracted, and diffracted. As we shall see later, they obey the principle of superposition and produce interference patterns. Because they are transverse waves, they can, in addition, be polarised (see Section 6.2). In a vacuum all electromagnetic waves travel at the same speed, $3.00 \times 10^8 \, \text{m s}^{-1}$.

The complete electromagnetic spectrum is divided into a series of regions based on the properties of electromagnetic waves in these regions, as illustrated in Figure 6.15. It should be noted that there is no clear boundary between regions.

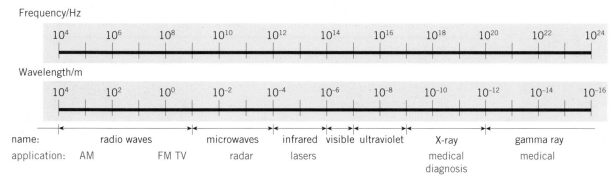

Figure 6.15 The electromagnetic spectrum

Figure 6.15 indicates some of the uses of electromagnetic waves. Most people are familiar with X-rays for medical diagnosis. Gamma radiation is also used for medical diagnosis. Since gamma radiation has high energy and penetration, it is also used in radiation therapy for the treatment of cancers and for sterilisation.

Infrared lasers are of particular importance in communications. Infrared light is not absorbed much by pure glass and is thus used for the transfer of information in optic fibres. The infrared source is a laser because this can be switched on and off very rapidly.

Ultraviolet (UV) light occupies the region of the spectrum between visible light and X-rays. This region is divided into three parts, according to wavelength.

UV-A occupies the region from 400 nm to 320 nm

UV-B is from 320 nm to 280 nm

UV-C is below 280 nm

Approximately 99% of the UV light from the Sun that reaches the Earth's surface is UV-A. A common form of UV lamp is the mercury vapour lamp: 86% of the UV light from such a lamp has a wavelength of 254 nm and is UV-C.

UV light is used in fluorescent lamps. A phosphorescent coating on the inside of the tube converts the UV light to visible light. Other applications of UV light include UV watermarks for security documents, chemical markers and the analysis of minerals. It is used for the curing of adhesives and coatings since it can cause polymerisation and hence hardening. UV light causes chain degradation in polymers such as poly(propylene). It also affects many pigments and dyes, causing changes in colour.

Some properties of UV light that are specific to wavelength and which affect living organisms are shown in Table 6.1.

Table 6.1 Properties of UV light

biological effect	UV-A 400–320 nm	UV-B 320–280 nm	UV-C below 280 nm
detected by some birds, reptiles and insects; partly responsible for generation of the ozone layer	✓		
phototherapy for skin conditions (e.g. psoriasis, eczema)	✓	✓	
sterilisation (e.g. pasteurising fruit juices, disinfecting drinking water)			✓
production of vitamin D in skin		✓	
destroys vitamin A in skin	✓	✓	
damage to collagen of fibres leading to accelerated ageing of skin	✓	✓	✓
produces sunburn		✓	
possible cause of skin cancers	✓	✓	

It can be seen from Table 6.1 that over-exposure to UV light may result in acute (e.g. sunburn) and chronic (e.g. cancer) skin conditions in humans. For anyone who is exposed to excessive amounts of sunlight, a sunscreen product that protects the skin from UV light is very necessary.

Sunscreen products have an SPF rating that indicates the level of protection applicable to UV-B (responsible for sunburn) but not UV-A. Such products may also contain titanium dioxide or zinc oxide to provide some protection against UV-A.

Section 6.1 Summary

- A progressive wave travels outwards from the source, carrying energy but without transferring matter.
- In a transverse wave, the oscillations are at right angles to the direction in which the wave carries energy.
- In a longitudinal wave, the oscillations are along the direction in which the wave carries energy.
- The intensity of a wave is the energy passing through unit area per unit time. Intensity is proportional to the square of the amplitude (and to the square of the frequency).
- The speed v, frequency f and wavelength λ of a wave are related by:
  $v = f\lambda$
- Properties of wave motion (reflection, refraction and diffraction) can be observed in a ripple tank.
- All wavelengths of electromagnetic radiation have the same speed $c = 3.00 \times 10^8 \, \text{m s}^{-1}$ in a vacuum.

Section 6.1 Questions

1 A certain sound wave in air has a speed $340 \, \text{m s}^{-1}$ and wavelength $1.7 \, \text{m}$. For this wave, calculate:
   (a) the frequency,
   (b) the period.

2 The speed of electromagnetic waves in vacuum (or air) is $3.0 \times 10^8 \, \text{m s}^{-1}$.
   (a) The visible spectrum extends from a wavelength of 400 nm (blue light) to 700 nm (red light). Calculate the range of frequencies of visible light.
   (b) A typical frequency for v.h.f. television transmission is 250 MHz. Calculate the corresponding wavelength.

3 Two waves travel with the same speed and have the same amplitude, but the first has twice the wavelength of the second. Calculate the ratio of the intensities transmitted by the waves.

4 A student stands at a distance of 5.0 m from a point source of sound, which is radiating uniformly in all directions. The intensity of the sound wave at her ear is $6.3 \times 10^{-6} \, \text{W m}^{-2}$.
   (a) The receiving area of the student's ear canal is $1.5 \, \text{cm}^2$. Calculate how much energy passes into her ear in one minute.
   (b) The student moves to a point 1.8 m from the source. Calculate the new intensity of the sound.

6.2 Polarisation of waves

At the end of Section 6.2 you should be able to:
- explain what is meant by plane-polarised waves
- explain that polarisation is a phenomenon associated with transverse waves only and understand the polarisation of electromagnetic waves
- state that light is partially polarised on reflection
- recall and apply Malus' law.

Think about generating waves in a rope by moving your hand holding the stretched rope up and down, or from side to side. The transverse vibrations of the rope will be in just one plane – the vertical plane if your hand is moving up and down, or the horizontal plane if it moves from side to side. The vibrations are said to be **plane-polarised** in either a vertical plane or a horizontal plane. However, if the rope passes through a vertical slit, then only if you move your hand up and down will the wave pass through the slit. This is illustrated in Figure 6.16. If the rope passes through a horizontal slit, only waves generated by a side-to-side motion of the hand will be transmitted. If the rope passes through a slit of one orientation followed by one of the other orientation, the wave is totally blocked whether it was initially in the vertical or the horizontal plane.

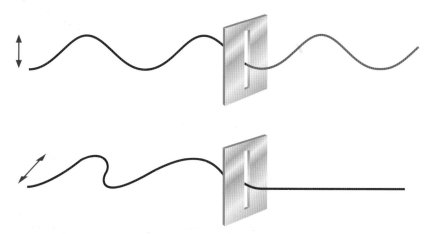

Figure 6.16 Transverse waves on a rope are plane-polarised.

The condition for a wave to be plane-polarised is for the vibrations to be in just one direction normal to the direction in which the wave is travelling.

Clearly, polarisation can apply only to transverse waves. Longitudinal waves vibrate along the direction of wave travel, and whatever the orientation of the slit, it would make no difference to the transmission of the waves.

The fact that light can be polarised was understood only in the early 1800s. This was a most important discovery. It showed that light is a transverse wave motion, and opened the way, 50 years later, to Maxwell's theory of light as electromagnetic radiation. Maxwell described light in terms of oscillating electric and magnetic fields, at right angles to each other and at right angles to the direction of travel of the wave energy (see Figure 6.14). When we talk about the direction of polarisation of a light wave, we refer to the direction of the electric field component of the electromagnetic wave.

The Sun and domestic light bulbs emit **unpolarised** light; that is, the vibrations take place in many directions at once, instead of in the single plane associated with plane-polarised radiation (Figure 6.17).

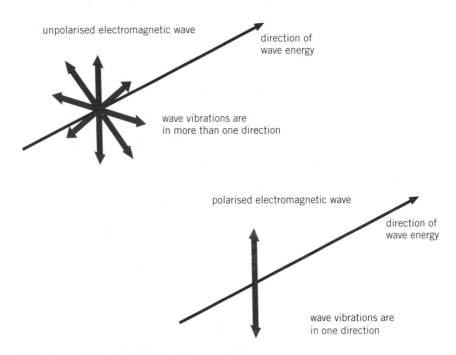

Figure 6.17 Unpolarised and polarised electromagnetic waves

Polarising light waves

Some transparent materials, such as a Polaroid sheet, allow vibrations to pass through in one direction only. A Polaroid sheet contains long chains of organic molecules aligned parallel to each other. When unpolarised light arrives at the sheet, the component of the electric field of the incident radiation which is parallel to the molecules is strongly absorbed, whereas radiation with its electric field perpendicular to the molecules is transmitted through the sheet. The Polaroid sheet acts as a **polariser**, producing plane-polarised light from light that was originally unpolarised. Figure 6.18 illustrates unpolarised light entering the polariser, and polarised light leaving it.

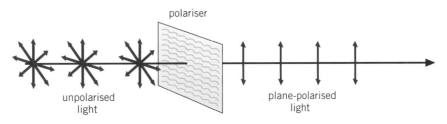

Figure 6.18 Action of a polariser

If you try to view plane-polarised light through a second sheet of Polaroid which is placed so that its polarising direction is at right angles to the polarising direction of the first sheet, it will be found that no light is transmitted. In this arrangement, the Polaroids are said to be crossed. The second Polaroid sheet is acting as an **analyser**. If the two Polaroids have their polarising directions parallel, then plane-polarised light from the first Polaroid can pass through the second. These two situations are illustrated in Figure 6.19. Although the action of the Polaroid sheet is not that of a simple slit, the arrangement of the crossed Polaroids has the same effect as the crossed slits in the rope-and-slits experiment (Figure 6.16).

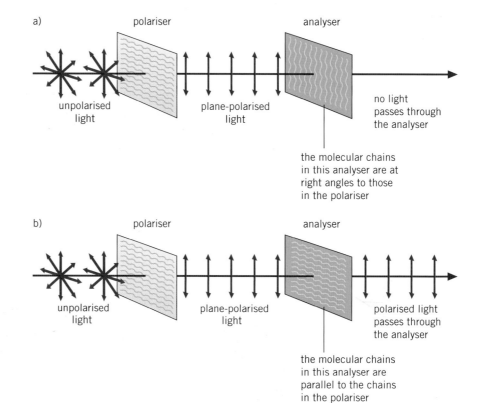

Figure 6.19 Polariser and analyser in a) crossed and b) parallel situations

Applications of polarisation

See the light but not the glare!

A very familiar application of the polarisation of light arises in the use of Polaroid sunglasses. Light reflected from a non-metallic surface, such as the smooth surface of a swimming-pool, is partially polarised parallel to the surface. Wearing sunglasses with Polaroid lenses, arranged with the polarising direction vertical, will prevent the transmission of much of the light reflected from the water. Reduction of the glare of the reflected light allows you to see objects more clearly. Fishermen find that Polaroid sunglasses eliminate glare from the surface of a stream or pond, so that they can see beneath the water more easily (Figure 6.20). Similarly, a polarising filter placed in front of a camera lens will eliminate glare, thus preventing 'hot spots' in photographs.

Figure 6.20 Wearing Polaroid sunglasses helps fishermen catch their prey!

Stress analysis

Figure 6.21 Stress analysis by polarised light

When polarised light passes through some transparent materials, the plane of polarisation is rotated. If the material is put under stress, the amount of stress affects the degree of rotation. If the incident polarised light is white, each of its component colours is rotated by a different amount, creating a pattern of coloured fringes when viewed through the analyser Polaroid. Engineers make use of this phenomenon by making models of structural components in a suitable transparent material, such as Perspex, stressing the model, and examining the pattern of fringes to identify regions of high stress. Where possible, the structure is re-designed to remove these danger points. Figure 6.19 illustrates a Perspex model being examined between crossed Polaroids.

Poor TV reception

In hilly areas television reception can be poor. To correct this problem the signal is often boosted by a local relay station before being re-transmitted. Sometimes this introduces further difficulties because of interference between the incident and the re-transmitted signal. This can be avoided if the main transmitter emits waves that are vertically polarised, and the relay station emits waves that are horizontally polarised. This is illustrated in Figure 6.22.

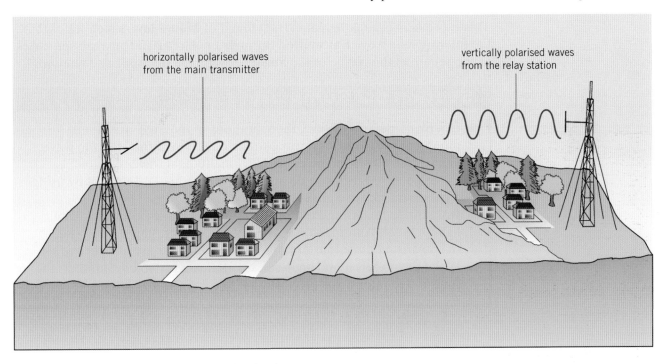

horizontally polarised waves from the main transmitter

vertically polarised waves from the relay station

Figure 6.22 Avoiding interference between television signals

Malus' law

Think about a wave that is incident on a polariser. The transmitted wave that is plane-polarised has an amplitude A and intensity I. Then we know that

$$I = kA^2$$

where k is a constant (see section 6.1).

An analyser is placed in the path of this wave so that the angle between the direction of polarisation of the polariser and the analyser is θ, as shown in Figure 6.23.

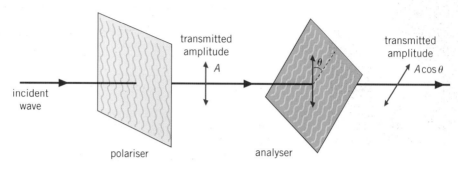

Figure 6.23 Demonstrating Malus' law

The component of the amplitude that will be transmitted by the analyser is $A \cos \theta$ and the transmitted intensity I_T will be given by

$$I_T = kA^2 \cos^2 \theta$$

This expression is a mathematical statement of **Malus' law**, which states that the intensity of a wave transmitted through a polariser and an analyser varies as the square of the cosine of the angle between the polariser and analyser.

Section 6.2 Summary

- In a plane-polarised wave, the vibrations of the wave are in one direction only, which is normal to the direction of travel of the wave.
- Transverse waves can be polarised; longitudinal waves cannot.
- Plane-polarised light can be produced from unpolarised light by using a polariser, such as a sheet of Polaroid.
- Rotating an analysing Polaroid in a beam of plane-polarised light prevents transmission of the polarised light. This occurs when the polarising directions of polariser and analyser are at right angles.
- When the polariser and analyser are at an angle θ to one another, the transmitted intensity is proportional to $\cos^2 \theta$ (Malus' law).

Section 6.2 Questions

1 You have been sold a pair of sunglasses, and are not sure whether they have Polaroid lenses. How could you find out?

2 The amplitude of a plane-polarised wave that has been transmitted through a polariser is A. The wave then passes through an analyser such that the angle between the polariser and the analyser is θ. Calculate, in terms of A, the intensity of the wave emerging from the analyser for angle θ equal to:
   (a) 0°,
   (b) 90°,
   (c) 60°.

6.3 Interference

At the end of Section 6.3 you should be able to:

- state and use the *principle of superposition* of waves
- apply graphical methods to illustrate the principle of superposition
- explain the terms *interference*, *coherence*, *path difference* and *phase difference*
- state what is meant by *constructive interference* and *destructive interference*
- describe experiments that demonstrate two-source interference
- describe the Young double-slit experiment
- select and use the equation $\lambda = ax/D$ for electromagnetic waves.

Any moment now the unsuspecting fisherman in Figure 6.24 is going to experience the effects of interference. The amplitude of oscillation of his boat will be significantly affected by the two approaching waves and their interaction when they reach his position.

Figure 6.24

If two or more waves overlap, the resultant displacement is the sum of the individual displacements. Remember that displacement is a vector quantity. The overlapping waves are said to **interfere**. This may lead to a resultant wave of either a larger or a smaller displacement than either of the two component waves.

Interference can be demonstrated in the ripple tank by using two point sources. Figure 6.26 shows such an interference pattern, and Figure 6.25 shows how it arises.

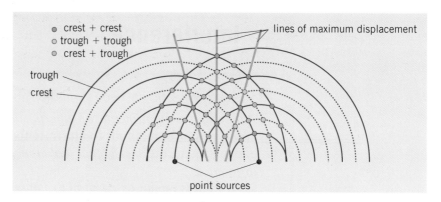

Figure 6.25 Two-source interference of circular waves in a ripple tank

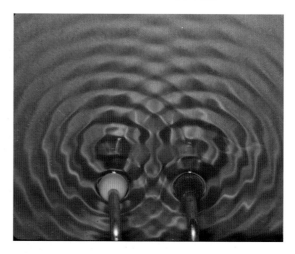

Figure 6.26 The resulting interference pattern

Figure 6.27 shows two waves arriving at a point at the same time. If they arrive **in phase** – that is, if their crests arrive at exactly the same time – they will interfere **constructively**. A resultant wave will be produced which has crests much higher than either of the two individual waves, and troughs which are much deeper. If the two incoming waves have the same frequency and equal amplitude A, the resultant wave produced by constructive interference has an amplitude of $2A$. The frequency of the resultant is the same as that of the incoming waves.

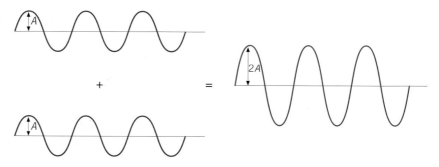

Figure 6.27 Constructive interference

If the two waves arrive **out of phase** (with a phase difference of π radians or 180°) – that is, if the peaks of one wave arrive at the same time as the troughs from the other – they will interfere **destructively**. The resultant wave will have a smaller amplitude. In the case shown in Figure 6.28, where the incoming waves have equal amplitude, the resultant wave has zero amplitude.

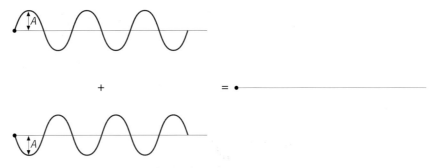

Figure 6.28 Destructive interference

This situation is an example of the **principle of superposition of waves**. The principle describes how waves, which meet at the same point in space, interact.

> The principle of superposition states that, when two or more waves meet at a point, the resultant displacement at that point is equal to the sum of the displacements of the individual waves at that point.

Because displacement is a vector, we must remember to add the individual displacements taking account of their directions. The principle applies to all types of wave.

If we consider the effect of superposition at a number of points in space, we can build up a pattern showing some areas where there is constructive interference, and hence a large wave disturbance, and other areas where the interference is destructive, and there is little or no wave disturbance.

Figure 6.29 illustrates the interference of waves from two point sources A and B. The point C is equidistant from A and B: a wave travelling to C from A moves through the same distance as a wave travelling to C from B. If the waves started out in phase at A and B, they will arrive in phase at C. They combine constructively, producing a maximum disturbance at C.

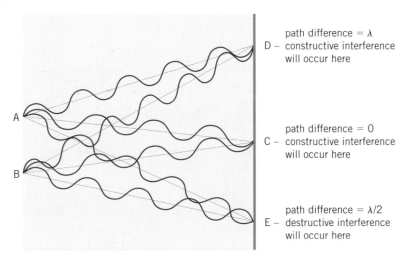

path difference = λ
D – constructive interference will occur here

path difference = 0
C – constructive interference will occur here

path difference = $\lambda/2$
E – destructive interference will occur here

Figure 6.29 Producing an interference pattern

149

At other places, such as D, the waves will have travelled different distances from the two sources. There is a **path difference** between the waves arriving at D. If this path difference is a whole number of wavelengths (λ, 2λ, 3λ, etc.) the waves arrive in phase and interfere constructively, producing maximum disturbance again. However, at places such as E the path difference is an odd number of half-wavelengths ($\lambda/2$, $3\lambda/2$, $5\lambda/2$, etc.). The waves arrive at E out of phase, and interference is destructive, producing a minimum resultant disturbance. This collection of maxima and minima produced by the superposition of overlapping waves is called an **interference pattern**. One is shown in Figure 6.26.

Producing an interference pattern with sound waves

Figure 6.30 shows an experimental arrangement to demonstrate interference with sound waves from two loudspeakers connected to the same signal generator and amplifier, so that each speaker produces a note of the same frequency. The demonstration is best carried out in the open air (on playing-fields, for example) to avoid reflections from walls, but it should be a windless day. Moving about in the space around the speakers, you pass through places where the waves interfere constructively and you can hear a loud sound. In places where the waves interfere destructively, the note is much quieter than elsewhere in the pattern.

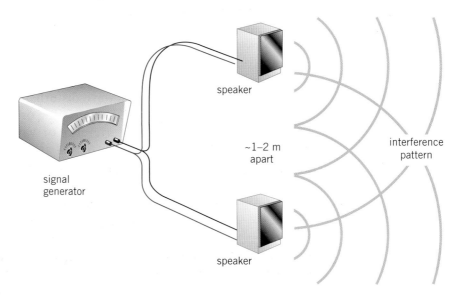

Figure 6.30 Demonstration of interference with sound waves

Producing an interference pattern with light waves

If you try to set up a demonstration with two separate light sources, such as car headlights, you will find that it is not possible to produce an observable interference pattern (Figure 6.31). A similar demonstration works with sound waves from two separate loudspeakers. What has gone wrong?

To produce an observable interference pattern, the two wave sources must have the same **single frequency**, not a mixture of frequencies as is the case for

Figure 6.31 Failure of an interference demonstration with light

light from car headlights. They must also have a **constant phase relationship**. In the sound experiment, the waves from the two loudspeakers have the same frequency and a constant phase relationship because the loudspeakers are connected to the same oscillator and amplifier. If the waves emitted from the speakers are in phase when the experiment begins, they stay in phase for the whole experiment.

> Wave sources which maintain a constant phase relationship are described as **coherent** sources.

This is illustrated in Figure 6.32.

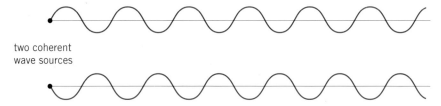

Figure 6.32 Coherent wave trains

Light is emitted from sources as a series of pulses or packets of energy. These pulses last for a very short time, about a nanosecond (10^{-9}s). Between each pulse there is an abrupt change in the phase of the waves. Waves from two separate sources may be in phase at one instant, but out of phase in the next nanosecond. The human eye cannot cope with such rapid changes, so the pattern is not observable. So separate light sources, even of the same frequency, produce incoherent waves (Figure 6.33).

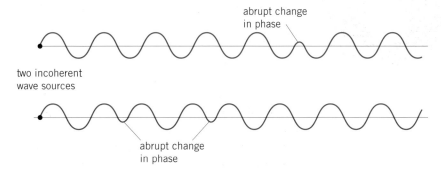

Figure 6.33 Incoherent light trains

To obtain observable interference patterns, it is not essential for the amplitudes of the waves from the two sources to be the same. However, if the amplitudes are not equal, a completely dark fringe will never be obtained, and the contrast of the pattern is reduced.

Young's double-slit experiment

In 1801 Thomas Young demonstrated how light waves could produce an interference pattern. The experimental arrangement is shown in Figure 6.34 (not to scale).

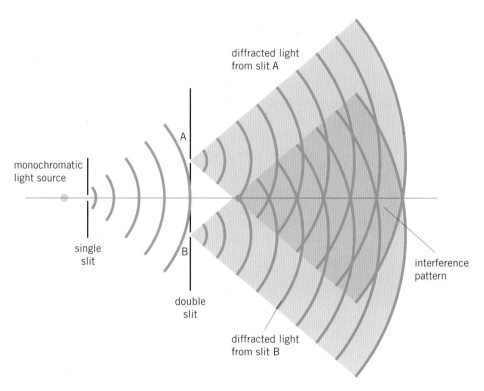

monochromatic
light source

single
slit

double
slit

A

B

diffracted light
from slit A

diffracted light
from slit B

interference
pattern

Figure 6.34 Young's double-slit experiment

Figure 6.35 Fringe pattern in Young's experiment

A monochromatic light source (a source of one colour, and hence one wavelength λ) is placed behind a single slit to create a small, well defined source of light. Light from this source is diffracted at the slit, producing two light sources at the double slit A and B. Because these two light sources originate from the same primary source, they are coherent and create a sustained and observable interference pattern, as seen in the photograph of the dark and bright interference fringes in Figure 6.35. Bright fringes are seen where constructive interference occurs – that is, where the path difference between the two diffracted waves from the sources A and B is $n\lambda$, where n is a whole number. Dark fringes are seen where destructive interference occurs. The condition for a dark fringe is that the path difference should be $(n + \frac{1}{2})\lambda$.

The distance x on the screen between successive bright fringes is called the fringe width. The fringe width is related to the wavelength λ of the light source by the equation

$$x = \frac{\lambda D}{a} \quad \text{or} \quad \lambda = \frac{ax}{D}$$

where D is the distance from the double slit to the screen and a is the distance between the centres of the slits. Note that, because the wavelength of light is so small (of the order of 10^{-7} m), to produce observable fringes D needs to be large and a as small as possible (Example, page 154). (This is another reason why you could never see an interference pattern from two sources such as car headlamps.)

Although Young's original double-slit experiment was carried out with light, the conditions for constructive and destructive interference apply for any two-source situation. The same formula applies for all types of wave, including sound waves, water waves and microwaves, provided that the fringes are detected at a distance of many wavelengths from the two sources.

The classical Young double-slit experiment provided evidence for a wave nature of light.

Determining the wavelength of laser light

This experiment is a modern version of Young's experiment, using laser light. A laser produces a narrow beam of coherent light. If this beam is incident on a double-slit arrangement, as shown in Figure 6.36 (not to scale), then interference fringes will be visible on a screen.

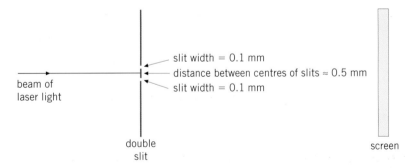

Figure 6.36 The double-slit experiment with laser light

The double-slit arrangement consists of two narrow parallel slits, such that their centres are separated by a distance of approximately 0.5 mm. Each slit has a width of approximately 0.1 mm. The screen in placed about 2.0 m from the double slit. The distance D between the double slit and the screen is measured with a metre rule. A travelling microscope is used to measure the slit separation a.

A number of fringes will be visible on the screen. The distance across as many fringes as possible is measured with a rule and divided by the number of fringes in order to find the fringe separation x. The wavelength λ is then determined using the formula

$$\lambda = \frac{ax}{D}$$

Determining the wavelength of microwaves

The double-slit arrangement is made using three sheets of aluminium so that each slit has a width of about 1 cm and the separation of the slits is about 4 cm. The arrangement of the apparatus is as shown in Figure 6.37 (not to scale).

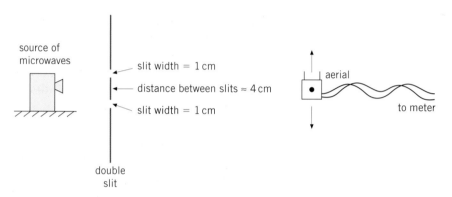

source of microwaves

slit width = 1 cm

distance between slits ≈ 4 cm

slit width = 1 cm

aerial

to meter

double slit

Figure 6.37 The double-slit experiment with microwaves

When the probe is moved along a line parallel to the plane of the slits, the signal strength passes through a series of maxima and minima. The distances a, x and D can be measured using a metre rule.

Example

Calculate the observed fringe width for a Young's double-slit experiment using light of wavelength 600 nm and slits 0.50 mm apart. The distance from the slits to the screen is 0.80 m.

Using $x = \lambda D/a$, $x = (600 \times 10^{-9} \times 0.80)/(0.50 \times 10^{-3}) = 9.6 \times 10^{-4}$ m
= **0.96 mm**

Now it's your turn

1 Calculate the wavelength of light which produces fringes of width 0.50 mm on a screen 60 cm from two slits 0.75 mm apart.

2 Microwaves of wavelength 50 mm create a fringe pattern 1.0 km from the aerials. Calculate the distance between the aerials if the fringe spacing is 80 cm.

Figure 6.38 White-light interference fringes

White-light fringes

If the two slits in Young's experiment are illuminated with white light, each of the different wavelengths making up the white light produces its own fringe pattern. At the centre of the pattern, where the path difference for all waves is zero, there will be a white maximum with a black fringe on each side (Figure 6.38). Thereafter, the maxima and minima of the different colours overlap in such a way as to produce a pattern of coloured fringes. Only a few will be visible; a short distance from the centre so many wavelengths overlap that they combine to produce what is effectively white light again.

Section 6.3 Summary

- The principle of superposition of waves states that when waves meet at the same point in space, the resultant displacement is given by the sum of the displacements of the individual waves.
- Constructive interference is obtained when the waves that meet are completely in phase, so that the resultant wave is of greater amplitude than any of its constituents.
- Destructive interference is obtained when the waves that meet are completely out of phase.
- To produce a sustained and observable interference pattern the sources must be coherent (have a constant phase relationship).
- Coherent sources have a constant phase difference between them.
- Young's double-slit experiment:
  - condition for constructive interference: path difference = $n\lambda$
  - condition for destructive interference: path difference = $(n + \frac{1}{2})\lambda$
  - fringe width $x = \lambda D/a$, where a is the separation of the source slits and D is the distance of the screen from the slits.

Section 6.3 Questions

1 Compare a two-source experiment to demonstrate the interference of sound waves with a Young's double-slit experiment using light. What are the similarities and differences between the two experiments?

2 (a) Explain the term *coherence* as applied to waves from two sources.
  (b) Describe how you would produce two coherent sources of light.
  (c) A double-slit interference pattern is produced using slits separated by 0.45 mm, illuminated with light of wavelength 633 nm from a laser. The pattern is projected on to a wall 2.50 m from the slits. Calculate the fringe separation.

3 Figure 6.39 shows the arrangement for obtaining interference fringes in a Young's double-slit experiment. Describe and explain what will be seen on the screen if the arrangement is altered in each of the following ways:
  (a) the slit separation a is halved,
  (b) the distance D from slits to screen is doubled,
  (c) the monochromatic light source is replaced with a white-light source,
  (d) a piece of Polaroid is placed in front of each slit, the polarising directions of the Polaroids being vertical for one slit and horizontal for the other.

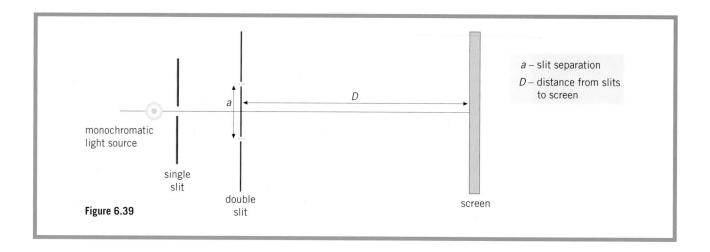

Figure 6.39

a – slit separation
D – distance from slits to screen

6.4 Diffraction

> At the end of Section 6.4 you should be able to:
> - describe the use of a diffraction grating to determine the wavelength of light
> - select and use the equation $d \sin \theta = n\lambda$
> - explain the advantages of using multiple slits in an experiment to find the wavelength of light.

Although we often hear the statement 'light travels in straight lines', there are occasions when this appears not to be the case. Newton tried to explain the fact that when light travels through an aperture, or passes the edge of an obstacle, it deviates from the straight-on direction and appears to spread out. We have seen from the ripple tank demonstration (Figure 6.13) that water waves spread out when they pass through an aperture. This effect is called **diffraction.** The fact that light undergoes diffraction is powerful evidence that light has wave properties. Newton's attempt to explain diffraction was not, in fact, based on a wave theory of light. The Dutch scientist Christian Huygens, a contemporary of Newton, favoured the wave theory, and used it to account for reflection, refraction and diffraction. (It was not until 1815 that the French scientist Augustin Fresnel developed the wave theory of light so as to explain diffraction in detail.)

The experiment illustrated in Figure 6.13 showed that the degree to which waves are diffracted depends upon the size of the obstacle or aperture and the wavelength of the wave. The greatest effects occur when the wavelength is about the same size as the aperture. The wavelength of light is very small (green light has wavelength 5×10^{-7} m), and therefore diffraction effects can be difficult to detect.

Huygens' explanation of diffraction

If we let a single drop of water fall into a ripple tank, it will create a circular wavefront which will spread outwards from the disturbance (Figure 6.10). Huygens put forward a wave theory of light which was based on the way in which circular wavefronts advance. He suggested that at any instant all points on a wavefront could be regarded as secondary disturbances, giving rise to their own outward-spreading circular wavelets. The envelope, or tangent curve, of the wavefronts produced by the secondary sources gives the new position of the original wavefront. This construction is illustrated in Figure 6.40 for a circular wavefront.

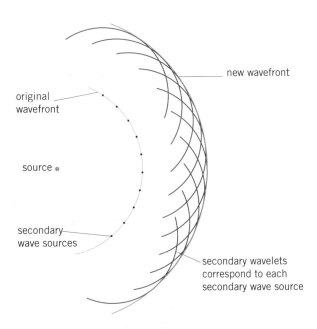

Figure 6.40 Huygens' construction for a circular wavefront

Now think about a plane (straight) wavefront. If the wavefront is restricted in any way, for example by passing through an aperture in the form of a slit, some of the wavelets making up the wavefront are removed, causing the edges of the wavefront to be curved. If the wavelength is small compared with the size of the aperture, the wavefronts which pass through the aperture show curvature only at their ends, and the diffraction effect is relatively small. If the aperture is comparable with the wavelength, the diffracted wavefronts become circular, centred on the slit. Note that there is no change of wavelength of diffraction. This effect is illustrated in Figure 6.13.

Figure 6.41 shows the diffraction pattern created by a single slit illuminated by monochromatic light. The central region of the pattern is a broad, bright area with narrow, dark fringes on either side. Beyond these is a further succession of bright and dark areas. The bright areas become less and less intense as we move away from the centre.

Figure 6.41 Diffraction of light at a single slit

The diffraction grating

A **diffraction grating** is a plate on which there is a very large number of parallel, identical, very closely spaced slits. If monochromatic light is incident on this plate, a pattern of narrow bright fringes is produced (Figure 6.42).

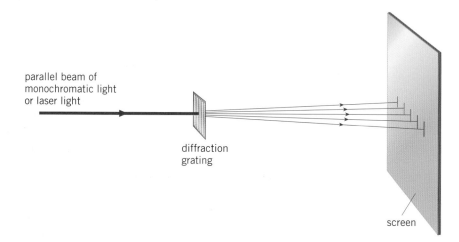

parallel beam of monochromatic light or laser light

diffraction grating

screen

Figure 6.42 Arrangement for obtaining a fringe pattern with a diffraction grating

Although the device is called a *diffraction* grating, we shall use straightforward superposition and interference ideas in obtaining an expression for the angles at which the maxima of intensity are obtained.

Figure 6.43 shows a parallel beam of light incident normally on a diffraction grating in which the spacing between adjacent slits is *d*. Consider first rays 1 and 2 which are incident on adjacent slits. The path difference between these rays when they emerge at an angle θ is $d \sin \theta$. To obtain constructive interference in this direction from these two rays, the condition is that the path difference should be an integral number of wavelengths. The path difference between rays 2 and 3, 3 and 4, and so on, will also be $d \sin \theta$. The condition for constructive interference is the same. Thus, the condition for a maximum of intensity at angle θ is

$$d \sin \theta = n\lambda$$

where λ is the wavelength of the monochromatic light used, and *n* is a whole number.

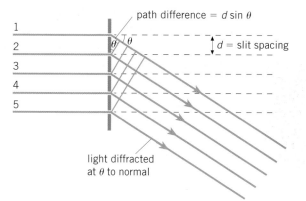

path difference = $d \sin \theta$

1
2
3
4
5

θ θ

d = slit spacing

light diffracted at θ to normal

Figure 6.43

When $n = 0$, $\sin \theta = 0$ and θ is also zero; this gives the straight-on direction, or what is called the zero-order maximum. When $n = 1$, we have the first-order diffraction maximum, and so on (Figure 6.44).

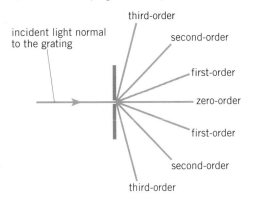

Figure 6.44 Maxima in the diffraction pattern of a diffraction grating

Determining the wavelength of laser light using a diffraction grating

The laser is arranged so that the laser beam is normal to the grating. The diffracted light is observed on a screen at least 1 m from the diffraction grating, as shown in Figure 6.45.

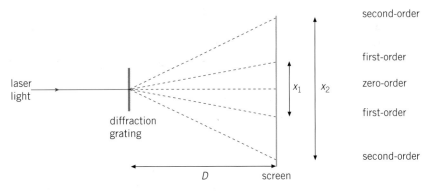

Figure 6.45 Determining the wavelength of laser light

The distance D between the grating and the screen is measured using a metre rule. The distance x_1 between the first-order images is measured. Similarly, the distance x_2 between the second-order images is found. The angle θ_1 for the first order diffracted light is given by

$$\tan \theta_1 = \frac{x_1}{2D}$$

The wavelength of the light is then found using the equation

$$\lambda = d \sin \theta_1$$

where d is the spacing of the slits in the grating. Similarly, for the second order,

$$\tan \theta_2 = \frac{x_2}{2D}$$

and

$$\lambda = \tfrac{1}{2}d \sin \theta_2$$

The average value for λ can then be found.

Determining the wavelength of the light from an LED using a diffraction grating

The light from the LED is not as intense as that from the laser and the light from the LED may be viewed directly. The arrangement of the apparatus is shown in Figure 6.46.

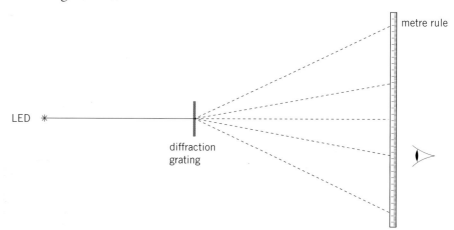

Figure 6.46 Determining the wavelength of light from an LED

The readings on the metre rule for the various orders of diffracted light are taken. The theory is the same as that for laser light.

Example

Monochromatic light is incident normally on a grating with 7.00×10^5 lines per metre. A second-order maximum is observed at an angle of diffraction of 40.0°. Calculate the wavelength of the incident light.

The slits on a diffraction grating are created by drawing parallel lines on the surface of the plate. The relationship between the slit spacing d and the number N of lines per metre is $d = 1/N$. For this grating, $d = 1/7.00 \times 10^5 = 1.43 \times 10^{-6}$ m. Using $n\lambda = d \sin \theta$, $\lambda = (d/n) \sin \theta = (1.43 \times 10^{-6}/2) \sin 40.0° = $ **460 nm**.

Now it's your turn

1 Monochromatic light is incident normally on a grating with 5.00×10^5 lines per metre. A third-order maximum is observed at an angle of diffraction of 78.0°. Calculate the wavelength of the incident light.

2 Light of wavelength 5.90×10^{-7} m is incident normally on a diffraction grating with 8.00×10^5 lines per metre. Calculate the diffraction angles of the first- and second-order diffraction images.

3 Light of wavelength 590 nm is incident normally on a grating with spacing 1.67×10^{-6} m. How many orders of diffraction maxima can be obtained?

The diffraction grating with white light

If white light is incident on a diffraction grating, each wavelength λ making up the white light is diffracted by a different amount, as described by the equation $n\lambda = d \sin \theta$. Red light, because it has the longest wavelength in the visible spectrum, is diffracted through the largest angle. Blue light has the shortest wavelength, and is diffracted the least. Thus, the white light is split into its component colours, producing a continuous spectrum (Figure 6.47). The spectrum is repeated in the different orders of the diffraction pattern. Depending on the grating spacing, there may be some overlapping of different orders. For example, the red component of the first-order image may overlap with the blue end of the second-order spectrum.

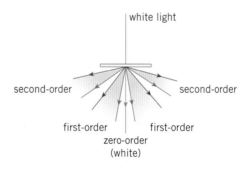

Figure 6.47 Production of the spectrum of white light with a diffraction grating

An important use of the diffraction grating is in a **spectrometer**, a piece of apparatus used to investigate spectra. By measuring the angle at which a particular diffracted image appears, the wavelength of the light producing that image may be determined.

Although the double-slit experiment is of historical importance, the diffraction grating has certain advantages for the determination of wavelength. First, because there are multiple slits, the images are brighter. Second, because the slits are closer together, the angles of diffraction are larger, giving greater accuracy of measurement.

Section 6.4 Summary

- Diffraction is the spreading out of waves after passing through an aperture or meeting an obstacle. It is most obvious when the size of the aperture and the wavelength of the wave are approximately the same.
- The condition for a diffraction maximum in a diffraction grating pattern is $d \sin \theta = n\lambda$, where d is the grating spacing, θ is the angle at which the diffraction maximum is observed, n is an integer (the order of the image), and λ is the wavelength of the light.

Section 6.4 Questions

1 Blue and red light, with wavelengths 450 nm and 650 nm respectively, is incident normally on a diffraction grating which has 4.0×10^5 lines per metre.

(a) Calculate the grating spacing.

(b) Calculate the angle between the second-order maxima for these wavelengths.

(c) For each wavelength, find the maximum order that can be observed.

2 Discuss any difference between the interference patterns formed by (a) two parallel slits 1 μm apart, (b) a diffraction grating with grating spacing 1 μm, when illuminated with monochromatic light.

6.5 Stationary waves

At the end of Section 6.4 you should be able to:

- explain the formation of stationary (standing) waves using graphical methods
- describe the similarities and differences between progressive and stationary waves
- define the terms *nodes* and *antinodes*
- describe experiments that demonstrate stationary waves
- determine the standing wave patterns for a stretched string and for air columns
- relate the separation of nodes and/or antinodes to wavelength
- define and use the terms *fundamental mode* and *harmonics*
- determine the speed of sound in air using stationary waves in a pipe closed at one end.

Figure 6.48 Cello being bowed

The notes we hear from a cello are created by the vibrations of its strings (Figure 6.48). The wave patterns on the vibrating strings are called **stationary waves** (or **standing waves**). The waves in the air which carry the sound to our ears transfer energy, and so are called **progressive waves**.

Figure 6.49 shows a single transverse pulse travelling along a 'slinky' spring. The pulse is reflected when it reaches the fixed end. If a second pulse is sent along the slinky (Figure 6.50), the reflected pulse will pass through the outward-going pulse, creating a new pulse shape. Interference will take place between the outward and reflected pulses.

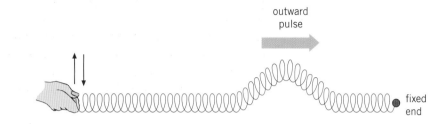

Figure 6.49 Single transverse pulse travelling along a slinky

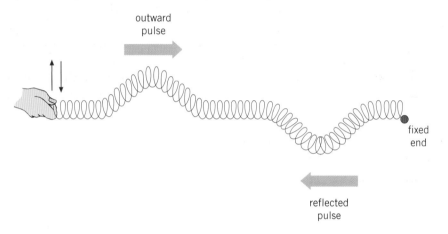

Figure 6.50 Reflected pulse about to meet an outward-going pulse

If the interval between outward pulses is reduced, a progressive wave is generated. When the wave reaches the fixed end, it is reflected. We now have two progressive waves of equal frequency and amplitude travelling in opposite directions on the same spring. The waves interfere, producing a wave pattern (Figure 6.51) in which the crests and troughs do not move. This pattern is called a **stationary** or **standing wave**, because it does not move. A stationary wave is the result of interference between two waves of equal frequency and amplitude, travelling along the same line with the same speed but in opposite directions.

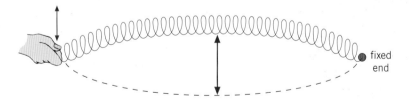

Figure 6.51 A stationary wave is created when two waves travelling in opposite directions interfere.

Stationary waves on strings

If a string is plucked and allowed to vibrate freely, there are certain frequencies at which it will vibrate. The amplitude of vibration at these frequencies is large. This is known as a **resonance** effect.

It is possible to investigate stationary waves in a more controlled manner using a length of string under tension and a vibrator driven by a signal generator. As the frequency of the vibrator is changed, different standing wave patterns are formed. Some of these are shown in Figure 6.52.

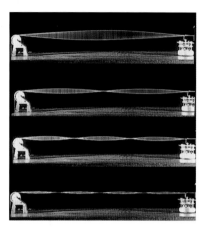

Figure 6.52 First four modes of vibration of a string

Figure 6.53 shows the simplest way in which a stretched string can vibrate. The wave pattern has a single loop. This is called the **fundamental mode** of vibration, or the **first harmonic**. At the ends of the string, there is no vibration. These points are called **nodes**. At the centre of the string, the amplitude is a maximum. A point of maximum amplitude is called an **antinode**. Nodes and antinodes do not move along the string.

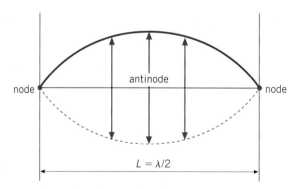

Figure 6.53 Fundamental mode of vibration of a stretched string

The wavelength λ of the standing wave in the fundamental mode is $2L$. From the wave equation $c = f\lambda$, the frequency f_1 of the note produced by the string vibrating in its fundamental mode is given by $f_1 = c/2L$, where c is the speed of the progressive waves which have interfered to produce the stationary wave.

Figure 6.54 shows the second mode of vibration of the string. The stationary wave pattern has two loops. This mode is sometimes called the **first overtone**, or the **second harmonic** (don't be confused!). The wavelength of this second mode is L. Applying the wave equation, the frequency f_2 is found to be c/L.

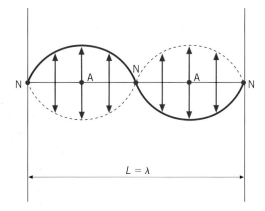

Figure 6.54 Second mode of vibration of a stretched string

Figure 6.55 shows the third mode (the second overtone, or third harmonic). This is a pattern with three loops. The wavelength is $2L/3$, and the frequency f_3 is $3c/2L$.

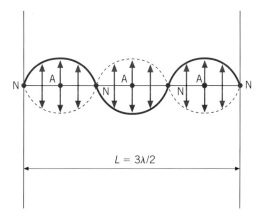

Figure 6.55 Third mode of vibration of a stretched string

The general expression for the frequency f_n of the nth mode (or the nth harmonic, or $(n - 1)$th overtone) is

$$f_n = \frac{cn}{2L} \quad n = 1, 2, 3, \ldots$$

The key features of a stationary wave pattern on a string, which distinguish it from a progressive wave, are as follows.

- The nodes and antinodes do not move along the string, whereas in a progressive wave, the crests and troughs do move along it.
- The amplitude of vibration varies with position along the string: it is zero at a node, and maximum at an antinode. In a progressive wave, all points have the same amplitude.
- Between adjacent nodes, all points of the stationary wave vibrate in phase. That is, all particles of the string are at their maximum displacement at the same instant. In a progressive wave, phase varies continuously along the wave.

Stationary waves explained by interference

Let us explain the formation of a stationary wave using the principle of superposition. The set of graphs in Figure 6.56 represents two progressive waves of equal amplitude and frequency travelling in opposite directions. The red wave is travelling from left to right, and the blue wave is going from right to left.

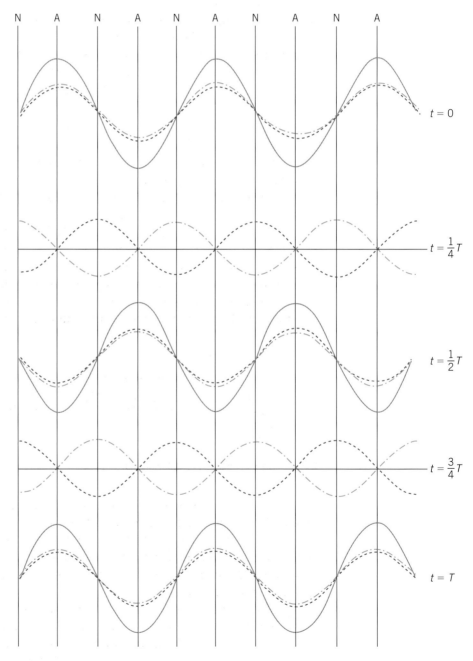

N   A   N   A   N   A   N   A   N   A

$t = 0$

$t = \frac{1}{4}T$

$t = \frac{1}{2}T$

$t = \frac{3}{4}T$

$t = T$

Figure 6.56 Formation of a stationary wave by superposition of two progressive waves travelling in opposite directions

The top graph catches the waves at an instant at which they are in phase. Superposition gives the purple curve, which has twice the amplitude of either of the progressive graphs. The second graph is the situation a quarter of a period (cycle) later, when the two progressive waves have each moved a quarter of a wavelength in opposite directions. This has brought them to a situation where the movement of one relative to the other is half a wavelength, so that the waves are exactly out of phase. The resultant, obtained by superposition, is zero everywhere. In the third graph, half a period from the start, the waves are again in phase, with maximum displacement for the resultant. The process continues through the fourth graph, showing the next out-of-phase situation, with zero displacement of the resultant everywhere. Finally, the fifth graph, one period on from the first, brings the waves into phase again.

We can see how there are some positions, the nodes N, where the displacement of the resultant is zero *throughout* the cycle. The displacement of the resultant at the antinodes A fluctuates from a maximum value when the two progressive waves are in phase to zero when they are out of phase.

Stationary waves in air

Figure 6.57 shows an experiment to demonstrate the formation of stationary waves in air. A fine, dry powder (such as cork dust or lycopodium powder) is sprinkled evenly along the transparent tube. A loudspeaker powered by a signal generator is placed at the open end. The frequency of the sound from the loudspeaker is gradually increased. At certain frequencies, the powder forms itself into evenly spaced heaps along the tube. A stationary wave has been set up in the air, caused by the interference of the sound wave from the loudspeaker and the wave reflected from the closed end of the tube. At nodes (positions of zero amplitude) there is no disturbance, and the powder can settle into heaps. At antinodes the disturbance is at a maximum, and the powder is dispersed.

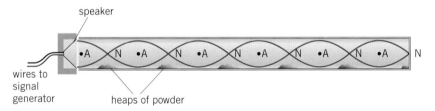

Figure 6.57 Demonstration of stationary waves in air

For stationary waves in a closed pipe, the air cannot move at the closed end, and so this must always be a node N. However, the open end is a position of maximum disturbance, and this is an antinode A. (In fact, the antinode is slightly outside the open end. The distance of the antinode from the end of the tube is called the end-correction. The value of the end-correction depends on the diameter of the tube.)

Figure 6.58 shows the simplest way in which the air in a pipe, closed at one end, can vibrate. Figure 6.58a illustrates the motion of some of the air particles in the tube. Their amplitude of vibration is zero at the closed end, and increases with distance up the tube to a maximum at the open end. This representation is tedious to draw, and Figure 6.58b is the conventional way of showing the

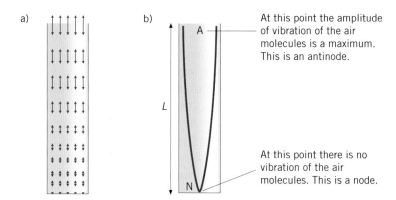

Figure 6.58 Fundamental mode of vibration of air in a closed pipe

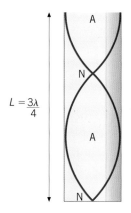

$L = \frac{3\lambda}{4}$

Figure 6.59 Second mode of vibration of air in a closed pipe

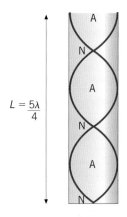

$L = \frac{5\lambda}{4}$

Figure 6.60 Third mode of vibration of air in a closed pipe

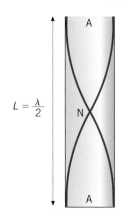

$L = \frac{\lambda}{2}$

Figure 6.61 Fundamental mode of vibration of air in an open pipe

amplitude of vibration: the amplitudes along the axis of the tube are plotted as a continuous curve. One danger of using diagrams like Figure 6.58b is that they give the impression that the sound wave is transverse rather than longitudinal. So be warned! The mode illustrated in Figure 6.58 is the fundamental mode (the first harmonic). The wavelength of this stationary wave (ignoring the end-correction) is $4L$, where L is the length of the pipe. Using the wave equation, the frequency f_1 of the fundamental mode is given by $f_1 = c/4L$, where c is the speed of the sound in air.

Other modes of vibration are possible. Figures 6.59 and 6.60 show the second mode (the first overtone, or second harmonic) and the third mode (the second overtone, or third harmonic), respectively. The corresponding wavelengths are $4L/3$ and $4L/5$, and the frequencies are $f_2 = 3c/4L$ and $f_3 = 5c/4L$.

The general expression for the frequency f_n of the nth mode of vibration of the air in the closed tube (the nth harmonic, or the $(n-1)$th overtone) is

$$f_n = \frac{(2n-1)c}{4L}$$

This is another example of resonance. The particular frequencies at which stationary waves are obtained in the pipe are the resonant frequencies of the pipe.

Standing waves can also be produced in a pipe that is open at both ends – an *open pipe*.

For a standing wave to be formed in an open pipe, there must be an antinode at each end of the pipe. The fundamental mode of vibration of air in an open pipe is illustrated in Figure 6.61.

The wavelength of this standing wave (ignoring the end-correction) is $2L$, where L is the length of the pipe. Using the wave equation, the frequency f_1 of this fundamental mode is given by

$$f_1 = \frac{c}{2L}$$

where c is the speed of the sound in air.

Other modes are possible. Figures 6.62 and 6.63 show the second mode (the first overtone or second harmonic) and the third mode (the second overtone or third harmonic), respectively.

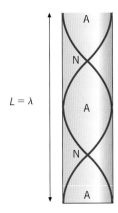

Figure 6.62 Second mode of vibration of air in an open pipe

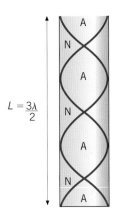

Figure 6.63 Third mode of vibration of air in an open pipe

The corresponding wavelengths are L and $2/3L$ and the frequencies are $f_2 = c/L$ and $f_2 = 3c/2L$.

The general expression for the frequency f_n of the nth mode of vibration of the air in an open pipe (the nth harmonic or the $(n-1)$th overtone is

$$f_n = \frac{nc}{2L}$$

Stationary waves using microwaves

Stationary waves can be demonstrated using microwaves. A source of microwaves faces a metal reflecting plate, as shown in Figure 6.64. A small detector is placed between source and reflector. The reflector is moved towards or away from the source until the signal picked up by the detector fluctuates regularly as it is moved slowly back and forth. The minima are nodes of the stationary wave pattern, and the maxima are antinodes. The distance moved by the detector between successive nodes is half the wavelength of the microwaves.

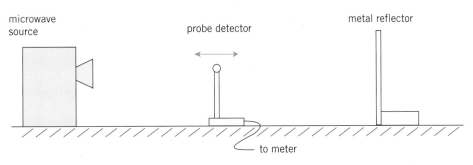

Figure 6.64 Using microwaves to demonstrate stationary waves

Measuring the speed of sound using stationary waves

The principle of resonance in a tube closed at one end can be used to measure the speed of sound in air. A glass tube is placed in a cylinder of water. By raising the tube, the length of the column of air can be increased. A vibrating tuning fork of known frequency f is held above the open end of the glass tube, causing the air in it to vibrate. The tube is gradually raised, increasing the length of the air column. At a certain position the note becomes much louder. This is known as the first position of resonance, and occurs when a stationary wave corresponding to the fundamental mode is established inside the tube. The length L_1 of the air column is noted. The tube is raised further until a second resonance position is found. This corresponds to the second mode of vibration. The length L_2 at this position is also noted. The two resonance positions are illustrated in Figure 6.65.

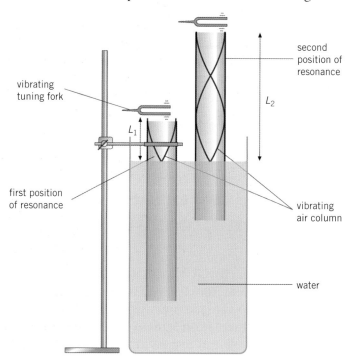

Figure 6.65 Speed of sound by the resonance tube method

At the first position of resonance, $\lambda/4 = L_1 + e$, where e is the end-correction of the tube (to allow for the fact that the antinode is slightly above the open end of the tube). At the second position of resonance, $3\lambda/4 = L_2 + e$. By subtracting these equations, we can eliminate e to give

$$\lambda/2 = L_2 - L_1$$

From the wave equation, the speed of sound c is given by $c = f\lambda$. Thus

$$c = 2f(L_2 - L_1)$$

Figure 6.66 illustrates a method of measuring the speed of sound using stationary waves in free air, rather than in a resonance tube. The signal generator and loudspeaker produce a note of known frequency f. The reflector is moved slowly back and forth until the trace on the oscilloscope has a

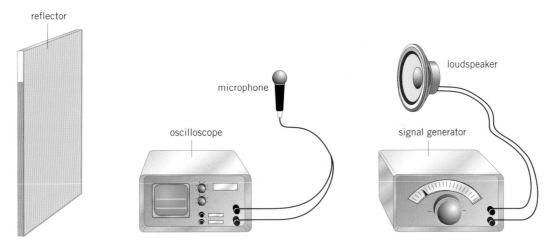

Figure 6.66 Speed of sound using stationary waves in free air

minimum amplitude. When this happens, a stationary wave has been set up with one of its nodes in the same position as the microphone. The microphone is now moved along the line between the loudspeaker and the reflector. The amplitude of the trace on the oscilloscope will increase to a maximum, and then decrease to a minimum. The microphone has been moved from one node, through an antinode, to the next node. The distance d between these positions is measured. We know that the distance between nodes is $\lambda/2$. The speed of sound can then be calculated using $c = f\lambda$, giving $c = 2fd$.

Examples

1 A string 75 cm long is fixed at one end. The other end is moved up and down with a frequency of 15 Hz. This frequency gives a stationary wave pattern with three complete loops on the string. Calculate the speed of the progressive waves which have interfered to produce the stationary wave.

The three-loop pattern corresponds to the situation where the length L of the string is $3\lambda/2$ (see Figure 6.55). The wavelength λ is thus $2 \times 0.75/3 = 0.50$ m. The frequency of the wave is 15 Hz, so by the wave equation $c = f\lambda = 15 \times 0.50 = \textbf{7.5 m s}^{-1}$.

2 Find the fundamental frequency and first two overtones for an organ pipe which is 0.17 m long and closed at one end. The speed of sound in air is 340 m s^{-1}.

The frequencies of the fundamental and first two overtones of a tube of length L, closed at one end, are $c/4L$, $3c/4L$ and $5c/4L$ (see Figures 6.58–6.60). The frequencies are thus $340/4 \times 0.17 = \textbf{500 Hz}$, $3 \times 340/4 \times 0.17 = \textbf{1500 Hz}$ and $5 \times 340/4 \times 0.17 = \textbf{2500 Hz}$.

Now it's your turn

1 A violin string vibrates with a fundamental frequency of 440 Hz. What are the frequencies of its first two overtones?

2 The speed of waves on a certain stretched string is $48\,\text{m s}^{-1}$. When the string is vibrated at frequency of 64 Hz, stationary waves are set up. Find the separation of successive nodes in the stationary wave pattern.

3 You can make an empty lemonade bottle resonate by blowing across the top. What fundamental frequency of vibration would you expect for a bottle 25 cm deep? The speed of sound in air is $340\,\text{m s}^{-1}$.

4 A certain organ pipe, closed at one end, can resonate at consecutive frequencies of 640 Hz, 896 Hz and 1152 Hz. Deduce its fundamental frequency.

Section 6.5 Summary

- A stationary wave is the result of interference between two progressive waves of equal frequency and amplitude travelling along the same line with the same speed, but in opposite directions.
- Points of zero amplitude on a stationary wave are called nodes; points of maximum amplitude are called antinodes.
- For stationary waves on a stretched string, frequency f_n of the nth mode is given by $f_n = cn/2L$, where c is the speed of progressive waves on the string and L is the length of the string.
- For stationary waves in air in a pipe closed at one end, frequency f_n of the nth mode is given by $f_n = (2_n - 1)c/4L$, where c is the speed of sound in air and L is the length of the pipe.
- For stationary waves in air in a pipe open at both ends, the frequency f_n of the nth mode is given by $f_n = nc/2L$, where c is the speed of sound in air and L is the length of the pipe.

Section 6.5 Questions

1 State and explain four ways in which stationary waves differ from progressive waves.

2 A source of sound of frequency 2000 Hz is placed in front of a flat wall. When a microphone is moved away from the source towards the wall, a series of maxima and minima are detected.
  (a) Explain what has happened to create these maxima and minima.
  (b) The speed of sound in air is $340\,\text{m s}^{-1}$. Calculate the distance between successive minima.

3 A string is stretched between two fixed supports separated by 1.20 m. Stationary waves are generated on the string. It is observed that two stationary wave frequencies are 180 Hz and 135 Hz; there is no resonant frequency between these two. Calculate:
  (a) the speed of progressive waves on the stretched string,
  (b) the lowest resonant frequency of the string.

Exam-style Questions

1 Assume that waves spread out uniformly in all directions from the epicentre of an earthquake. The intensity of a particular earthquake wave is measured as 5.0×10^6 W m^{-2} at a distance of 40 km from the epicentre. What is the intensity at a distance of only 2 km from the epicentre?

2 A string is stretched between two fixed supports 3.5 m apart. Stationary waves are generated by disturbing the string. One possible mode of vibration of the stationary waves is shown in Figure 6.67. The nodes and antinodes are labelled N and A respectively.

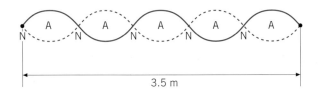

Figure 6.67

(a) Distinguish between a node and an antinode in a stationary wave.

(b) State the phase difference between the vibrations of particles of the string at any two neighbouring antinodes.

(c) Calculate the ratio of the frequency of the mode of vibration shown in Figure 6.67 to the frequency of the fundamental mode of vibration of the string.

(d) The frequency of the mode of vibration shown in Figure 6.67 is 160 Hz. Calculate the speed of the progressive waves which produced this stationary wave.

3 Explain why light will not pass through two sheets of Polaroid when they are arranged in contact with their polarising directions at right angles. What would happen if a third sheet of Polaroid, with its polarising direction at 45° to the polarising directions of the other two, were placed between them? (If you can get hold of Polaroid filters, it is worth trying this experiment, starting with thinking about how you would determine the polarising direction of each!)

4 A vibrating tuning fork of frequency 320 Hz is held over the open end of a resonance tube. The other end of the tube is immersed in water. The length of the air column is gradually increased until resonance first occurs. Taking the speed of sound in air as 340 m s^{-1}, calculate the length of the air column. (Neglect any end-correction.)

5 We can hear sounds round corners. We cannot see round corners. Both sound and light are waves. Explain why sound and light seem to behave differently.

6 Figure 6.68 shows a narrow beam of monochromatic laser light incident normally on a diffraction grating. The central bright spot is formed at O.

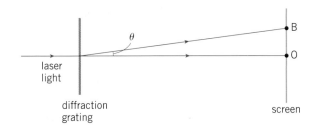

Figure 6.68

(a) Write down the relationship between the wavelength λ of the light and the angle θ for the first diffraction image formed at B. Identify any other symbol used.

(b) The screen is 1.1 m from the diffraction grating and the grating has 300 lines per mm. The laser light has wavelength 6.3×10^{-7} m. Find the distance OB from the central spot to the first bright image at B.

(c) The diffraction grating is now replaced by one which has 600 lines per mm. For this second grating, calculate the distance from the central spot to the first bright image.

7. Quantum physics

In the last chapter we spent a long time in describing properties of waves, and in emphasising experiments which showed that light is a wave motion. We now look at phenomena which, during the last hundred years, have completely revolutionised physics. A number of experiments around the 1900s showed that light behaves more like a stream of particles than as a wave. Later, it was shown experimentally that electrons, which had always been thought of as particles, could also behave as waves. This idea of *wave–particle duality* is one of the most important concepts of modern physics. We start by describing the photoelectric effect, the phenomenon that demonstrated that light had a particle nature. This is the work which Einstein first explained in 1905, and for which he received the 1921 Nobel Prize. We shall then describe the diffraction of electrons. Remember that diffraction is a characteristic of waves, yet electrons have always been regarded as particles!

7.1 The photoelectric effect

At the end of Section 7.1 you should be able to:
- describe and explain the phenomenon of the photoelectric effect
- explain that the photoelectric effect provides evidence for a particulate nature of electromagnetic radiation while phenomena such as interference and diffraction provide evidence for a wave nature.
- define and use the terms *work function* and *threshold frequency*
- select, explain and use Einstein's photoelectric equation $hf = \phi + (E_k)_{max}$
- state that a photon is a quantum of energy of electromagnetic radiation
- select and use the equations for photon energy $E = hf$ and $E = hc/\lambda$
- describe an experiment using LEDs to estimate the Planck constant using the equation $eV = hc/\lambda$.

Some of the electrons in a metal are free to move around in it. (It is these free electrons that form the electric current when a potential difference is applied across the ends of a metal wire.) However, to remove free electrons from a metal requires energy, because they are held in the metal by the electrostatic attraction of the positively charged nuclei. If an electron is to escape from the surface of a metal, work must be done on it. The electron must be given energy. When this energy is in the form of light energy, the phenomenon is called **photoelectric emission**.

Photoelectric emission is the release of electrons from the surface of a metal when electromagnetic radiation is incident on its surface.

Demonstration of photoelectric emission

A clean zinc plate is placed on the cap of a gold-leaf electroscope. The electroscope is then charged negatively, and the gold leaf deflects, proving that the zinc plate is charged. This is illustrated in Figure 7.1. If visible light of any colour is shone on to the plate, the leaf does not move. Even when the intensity (the brightness) of the light is increased, the leaf remains in its deflected position. However, when ultraviolet radiation is shone on the plate the leaf falls immediately, showing that it is losing negative charge. This means that electrons are being emitted from the zinc plate. These electrons are called **photoelectrons**. If the intensity of the ultraviolet radiation is increased, the leaf falls more quickly, showing that the rate of emission of electrons has increased.

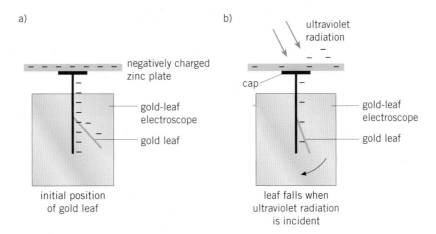

Figure 7.1 Demonstration of photoelectric emission

The difference between ultraviolet radiation and visible light is that ultraviolet radiation has a shorter wavelength and a higher frequency than visible light.

Further investigations with apparatus like this lead to the following conclusions:

■ If photoemission takes place, it does so instantaneously. There is no delay between illumination and emission.
■ Photoemission takes place only if the frequency of the incident radiation is above a certain minimum value called the **threshold frequency** f_0.

■ Different metals have different threshold frequencies.
■ Whether or not emission takes place depends only on whether the frequency of the radiation used is above the threshold for that surface. It does not depend on the intensify of the radiation.
■ For a given frequency, the rate of emission of photoelectrons is proportional to the intensity of the radiation.

Another experiment, using the apparatus shown in Figure 7.2, can be carried out to investigate the energies of the photoelectrons. If ultraviolet radiation of a fixed frequency (above the threshold) is shone on to the metal surface A, it emits photoelectrons. Some of these electrons travel from A to B. Current is detected using the microammeter. If a potential difference is applied between A and B, with B negative with respect to A, any electron going from A to B will gain potential energy as it moves against the electric field. The gain in potential energy is at the expense of the kinetic energy of the electron. That is,

loss in kinetic energy = gain in potential energy
= charge of electron × potential difference

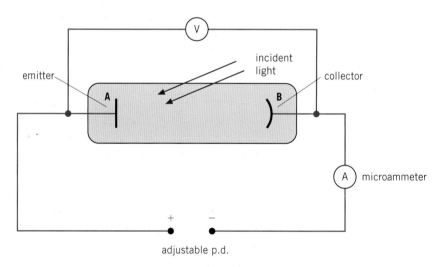

Figure 7.2 Experiment to measure the maximum kinetic energy of photoelectrons

If the voltage between A and B is gradually increased, the current registered on the microammeter decreases and eventually falls to zero. The minimum value of the potential difference necessary to stop the electron flow is known as the stopping potential. It measures the maximum kinetic energy with which the photoelectrons are emitted. The fact that there is a current in the microammeter at voltages less than the stopping potential indicates that there is range of kinetic energies for these electrons.

If the experiment is repeated with radiation of greater intensity but the same frequency, the maximum current in the microammeter increases, but the value of the stopping potential is unchanged.

The experiment can be repeated using ultraviolet radiation of different frequencies, measuring the stopping potential for each frequency. When the maximum kinetic energy of the photoelectrons is plotted against the frequency of the radiation, the graph of Figure 7.3 is obtained.

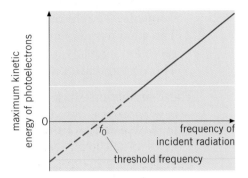

Figure 7.3 Graph of maximum kinetic energy of photoelectrons against frequency of radiation

The following conclusions are drawn from this experiment:

1 The photoelectrons have a range of kinetic energies, from zero up to some maximum value. If the frequency of the incident radiation is increased the maximum kinetic energy of the photoelectrons also increases.
2 For constant frequency of the incident radiation, the maximum kinetic energy is unaffected by the intensity of the radiation.
3 When the graph of Figure 7.3 is extrapolated to the point where the maximum kinetic energy of the photoelectrons is zero, the minimum frequency required to cause emission from the surface (the threshold frequency) may be found.

At the time when the photoelectric effect was first being studied, it was fully accepted that light is a wave motion. Evidence for this came from observed interference and diffraction effects. The conclusions of experiments on photoemission produced doubt as to whether light is a continuous wave. One of the main problems concerns the existence of a threshold frequency.

Classical wave theory predicts that when an electromagnetic wave (that is, light) interacts with an electron, the electron will absorb energy from it. So, if an electron absorbs enough energy it should be able to escape from the metal. Remember from Chapter 6 that the energy carried by a wave depends on its amplitude and its frequency. Thus, even if we have a low-frequency wave, its energy can be boosted by increasing the amplitude (that is, by increasing the brightness of the light). So, according to wave theory, we ought to be able to cause photoemission by using any frequency of light, provided we make it bright enough. Alternatively, we could use less bright light and shine it on the metal for a longer time, until enough energy to cause emission has been delivered. But this does not happen. The experiments we have described above showed conclusively that radiation of frequency below the threshold, no matter how intense or for how long it is used, does not produce photoelectrons. The classical wave theory of electromagnetic radiation leads to the following predictions:

1 Whether an electron is emitted or not should depend on the power of the incident wave; that is, on its intensity. A very intense wave, of any frequency, should cause photoemission.
2 The maximum kinetic energy of the photoelectrons should be greater if the radiation intensity is greater.
3 There is no reason why photoemission should be instantaneous.

These predictions, based on wave theory, do not match the observations. A new approach, based on an entirely new concept, the **quantum theory**, was used to explain these findings.

Einstein's theory of photoelectric emission

In 1901, the German physicist Max Planck had suggested that the energy carried by electromagnetic radiation might exist as discrete packets called **quanta**. The energy E carried in each quantum is given by

$$E = hf$$

where f is the frequency of the radiation and h is a constant called the Planck constant. The value of the Planck constant is 6.63×10^{-34} J s. Frequency f and wavelength λ are related by the equation $c = f\lambda$, where c is the speed of light. Thus energy E of each quantum $= hf = hc/\lambda$.

In 1905, Albert Einstein developed the theory of quantised energy to explain all the observations associated with photoelectric emission. He proposed that light radiation consists of a stream of energy packets called **photons**.

> A photon is the special name given to a quantum of energy when the energy is in the form of electromagnetic radiation.

When a photon interacts with an electron, it transfers all its energy to the electron. It is only possible for a single photon to interact with a single electron; the photon cannot share its energy between several electrons. This transfer of energy is instantaneous.

The photon theory of photoelectric emission is as follows. If the frequency of the incident radiation is less than the threshold frequency for the metal, the energy carried by each photon is insufficient for an electron to escape the surface of the metal. If the photon energy is insufficient for an electron to escape, it is converted to thermal energy in the metal.

> The minimum amount of energy necessary for an electron to escape from the surface is called the **work function energy** (or **work function**) ϕ.

Figure 7.4 Albert Einstein

Some values for the work function energy ϕ and threshold frequency f_0 of various metals are given in Table 7.1.

Table 7.1 Work function energies and threshold frequencies

metal	ϕ/J	ϕ/eV	f_0/Hz
sodium	3.8×10^{-19}	2.4	5.8×10^{14}
calcium	4.6×10^{-19}	2.9	7.0×10^{14}
zinc	5.8×10^{-19}	3.6	8.8×10^{14}
silver	6.8×10^{-19}	4.3	1.0×10^{15}
platinum	9.0×10^{-19}	5.6	1.4×10^{15}

Remember: The electronvolt is the energy gained by an electron when it moves through a potential difference of 1 volt. $1\,\text{eV} = 1.6 \times 10^{-19}\,\text{J}$.

If the frequency of the incident radiation is equal to the threshold frequency, the energy carried by each photon is just sufficient for electrons at the surface to escape. If the frequency of the incident radiation is greater than the threshold frequency, surface electrons will escape and have surplus energy in the form of kinetic energy. These electrons will have the maximum kinetic energy. If a photon interacts with an electron below the surface, some energy is used to take the electron to the surface, so that it is emitted with less than the maximum kinetic energy. This gives rise to a range of values of kinetic energy.

Einstein used the principle of conservation of energy to derive the photoelectric equation

photon energy = work function energy + maximum kinetic energy of photoelectron

or

$$hf = \phi + \tfrac{1}{2}m_e v_{max}^2$$

For radiation incident at the threshold frequency, $\tfrac{1}{2}m_e v_{max}^2 = 0$, so that $hf_0 = \phi$. The photoelectric equation can then be written as

$$hf = hf_0 + \tfrac{1}{2}m_e v_{max}^2$$

Determination of the Planck constant

The experiment described in Figure 7.2 enables a value to be determined for the Planck constant. The data collected in the experiment provides a graph showing the variation with frequency f of the maximum kinetic energy KE_{max} of the photoelectrons (see Figure 7.3).

As shown above, the equation for this graph is

photon energy = work function energy + maximum kinetic energy

The work function energy is a constant and therefore the equation may be written as

$$hf = constant + KE_{max}$$

or

$$f = \frac{constant}{h} + \frac{KE_{max}}{h}$$

This is the equation of the straight line of Figure 7.3. The gradient of this line is $1/h$.

An alternative means by which the Planck constant may be estimated is to use a light-emitting diode (LED), which emits monochromatic light when it is forward-biased. The LED gives off light only when the forward-biasing voltage provides enough energy for electrons to move to an excited state and then to de-excite giving out photons of wavelength λ. If the minimum forward-bias voltage for light emission is V, then the wavelength of this emitted light is given by

$$\frac{hc}{\lambda} = eV$$

where h is the Planck constant, c is the speed of light ($3.0 \times 10^8 \, m \, s^{-1}$) and e is the charge on an electron ($1.6 \times 10^{-19} \, C$).

An LED that emits monochromatic light of known wavelength is connected into the circuit shown in Figure 7.5.

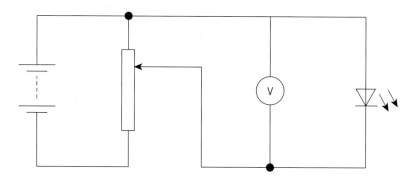

Figure 7.5 Circuit for the estimation of the Planck constant

The potential difference is gradually increased from zero until the LED just emits light. The potential difference V and the wavelength λ are noted.

The procedure is repeated for LEDs emitting different colours of light. A graph showing the variation with $1/\lambda$ of V is plotted, as shown in Figure 7.6.

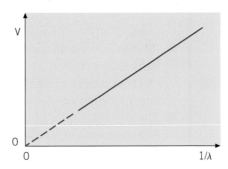

Figure 7.6 Graph for the estimation of the Planck constant

The gradient G of the graph is equal to hc/e and the Planck constant h is given by

$$h = \frac{Ge}{c}$$

where c is the speed of light and e is the charge on an electron.

Example

The work function energy of platinum is 9.0×10^{-19} J. Calculate:
(a) the threshold frequency for the emission of photoelectrons from platinum,
(b) the maximum kinetic energy of a photoelectron when radiation of frequency 2.0×10^{15} Hz is incident on a platinum surface.

(a) Using $hf_0 = \phi, f_0 = \phi/h$, so

$f_0 = 9.0 \times 10^{-19}/6.6 \times 10^{-34} = \mathbf{1.4 \times 10^{15}\,Hz}$

(b) Using $hf = hf_0 + \frac{1}{2}m_e v_{max}^2$, $hf - hf_0 = \frac{1}{2}m_e v_{max}^2$ and

$\frac{1}{2}m_e v_{max}^2 = 6.6 \times 10^{-34}\,(2.0 \times 10^{15} - 1.4 \times 10^{15}) = \mathbf{4.0 \times 10^{-19}\,J}$

Now it's your turn

1 The work function energy of silver is 4.3 eV. Show that the threshold frequency is about 1.0×10^{15} Hz.

2 Electromagnetic radiation of frequency 3.0×10^{15} Hz is incident on the surface of sodium metal. The emitted photoelectrons have a maximum kinetic energy of 1.6×10^{-18} J. Calculate the threshold frequency for photoemission from sodium.

Section 7.1 Summary

- Electrons may be emitted from metal surfaces if the metal is illuminated by electromagnetic radiation. This phenomenon is called photoelectric emission.
- Photoelectric emission cannot be explained by the wave theory of light. It is necessary to use the quantum theory, in which electromagnetic radiation is thought of as consisting of packets of energy called photons.
- The energy E of a photon is given by: $E = hf = hc/\lambda$, where h is the Planck constant, f is the frequency of the radiation, c is the speed of light and λ is the wavelength.
- The work function energy ϕ of a metal is the minimum energy needed to free an electron from the surface of the metal.
- The Einstein photoelectric equation is: $hf = \phi + \frac{1}{2}m_e v_{max}^2$
- The threshold frequency f_0 is given by: $hf_0 = \phi$

Section 7.1 Questions

1 Calculate the energy range, in eV, of photons in the visible region of the electro-magnetic spectrum (wavelengths 400 nm–700 nm).
(Planck constant $h = 6.6 \times 10^{-34}$ J s; speed of light $c = 3.0 \times 10^8$ m s^{-1}; electron charge $e = -1.6 \times 10^{-19}$ C)

2 The threshold frequency for photoemission from a certain metal is 8.8×10^{14} Hz.
(a) To what wavelength does this frequency correspond?
(b) Calculate the maximum kinetic energy of the emitted photoelectrons when the metal is illuminated with ultraviolet radiation of wavelength 240 nm.
(Planck constant $h = 6.6 \times 10^{-34}$ J s; speed of light $c = 3.0 \times 10^8$ m s^{-1})

3 In a stopping-potential experiment using ultra-violet radiation of wavelength 265 nm, the photocurrent from a certain metal is reduced to zero at an applied potential difference of 1.18 V. Calculate the work function energy of the metal.
(Planck constant $h = 6.63 \times 10^{-34}$ J s; speed of light $c = 3.00 \times 10^8$ m s^{-1}; electron charge $e = -1.60 \times 10^{-19}$ C)

7.2 Wave–particle duality

At the end of Section 7.2 you should be able to:
- explain electron diffraction as evidence for the wave nature of particles like electrons
- select and apply the de Broglie equation $\lambda = h/p$, $\lambda = h/mv$
- explain that the diffraction of electrons by matter can be used to determine the arrangement of atoms and the size of nuclei.

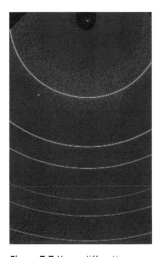

Figure 7.7 X-ray diffraction pattern of a metal foil

Figure 7.8 Electron diffraction pattern of a metal foil

If light waves can behave like particles (photons), perhaps moving particles can behave like waves?

If a beam of X-rays of a single wavelength is directed at a thin metal foil, a diffraction pattern is produced (Figure 7.7). This is a similar effect to the diffraction pattern produced when light passes through a diffraction grating (see Chapter 6). The foil contains many tiny crystals. The gaps between neighbouring planes of atoms in the crystals act as slits, creating a diffraction pattern.

If a beam of electrons is directed at a thin metal foil, a similar diffraction pattern is produced, as shown in Figure 7.8. The electrons, which we normally consider to be particles, are exhibiting a property we would normally associate with waves. Remember that, to observe diffraction, the wavelength of the radiation should be comparable with the size of the aperture. The separation of planes of atoms in crystals is of the order of 10^{-10} m. The fact that diffraction is observed with electrons suggest that they have a wavelength of about the same magnitude.

In 1924 the French physicist Louis de Broglie suggested that all moving particles have a wave-like nature. Using ideas based upon the quantum theory and Einstein's theory of relativity, he suggested that the momentum p of a particle and its associated wavelength λ are related by the equation

$$\lambda = \frac{h}{p}$$

where h is the Planck constant. λ is known as the **de Broglie wavelength**.

It can be seen that the de Broglie wavelength is dependent on momentum. Therefore, for electrons, the de Broglie wavelength decreases with increase of the potential difference through which the electrons are accelerated. By suitable choice of accelerating potential difference, the de Broglie wavelength can be selected so that electron diffraction may be used not only to determine the arrangement of atoms in crystals but also the size of atoms.

Example

Calculate the de Broglie wavelength of an electron travelling with a speed of 1.0×10^7 m s^{-1}.
(Planck constant $h = 6.6 \times 10^{-34}$ J s; electron mass $m_e = 9.1 \times 10^{-31}$ kg)

Using $\lambda = h/p$ and $p = mv$,
$\lambda = 6.6 \times 10^{-34}/9.1 \times 10^{-31} \times 1.0 \times 10^7 = \mathbf{7.3 \times 10^{-11}}$ **m**

Now it's your turn

1 Calculate the de Broglie wavelength of an electron travelling with a speed of 5.5×10^7 m s^{-1}.
(Planck constant $h = 6.6 \times 10^{-34}$ J s; electron mass $m_e = 9.1 \times 10^{-31}$ kg)

2 Calculate the de Broglie wavelength of an electron which has been accelerated from rest through a potential difference of 100 V.
(Planck constant $h = 6.6 \times 10^{-34}$ J s; electron mass $m_e = 9.1 \times 10^{-31}$ kg; electron charge $e = -1.6 \times 10^{-19}$ C)

Section 7.2 Questions

1 Calculate the de Broglie wavelength of an electron with energy 1.0 keV.
(Planck constant $h = 6.6 \times 10^{-34}$ J s; electron mass $m_e = 9.1 \times 10^{-31}$ kg; 1 eV $= 1.6 \times 10^{-19}$ J)

2 Calculate the speed of a neutron with de Broglie wavelength 1.5×10^{-10} m.
(Planck constant $h = 6.6 \times 10^{-34}$ J s; neutron mass $m_n = 1.7 \times 10^{-27}$ kg)

7.3 Emission spectra

At the end of Section 7.3 you should be able to:
■ explain how spectral lines are evidence for the existence of discrete energy levels in isolated atoms
■ describe the origin of emission and absorption line spectra
■ use the relationships $hf = E_2 - E_1$ and $hc/\lambda = E_2 - E_1$.

Continuous and line spectra

If white light from a tungsten filament lamp is passed through a prism, the light is dispersed into its component colours, as illustrated in Figure 7.9. The band of different colours is called a **continuous spectrum**. A continuous spectrum has all colours (and wavelengths) between two limits. In the case of white light, the colour and wavelength limits are violet (about 400 nm) and red (about 700 nm). Since this spectrum has been produced by the emission of light from the tungsten filament lamp, it is referred to as an **emission spectrum**. Finer detail of emission spectra than is obtained using a prism may be achieved using a diffraction grating.

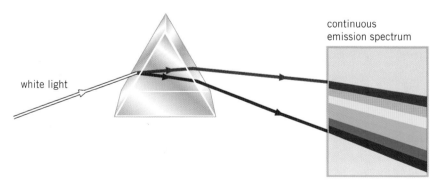

white light

continuous emission spectrum

Figure 7.9 Continuous spectrum of white light from a tungsten filament lamp

A discharge tube is a transparent tube containing a gas at low pressure. When a high potential difference is placed across two electrodes in the tube, light is emitted. Examination of the light with a diffraction grating shows that the emitted spectrum is no longer continuous, but consists of a number of bright lines (Figure 7.10).

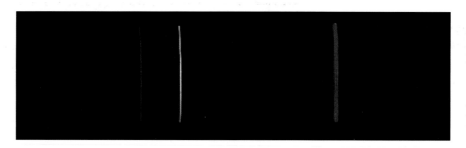

Figure 7.10 Line spectrum of hydrogen from a discharge tube

Such a spectrum is known as a **line spectrum**. It consists of a number of separate colours, each colour being seen as the image of the slit in front of the source. The wavelengths corresponding to the lines of the spectrum are characteristic of the gas which is in the discharge tube.

Electron energy levels in atoms

To explain how line spectra are produced we need to understand how electrons in atoms behave. Electrons in an atom can have only certain specific energies. These energies are called the **electron energy levels** of the atom. The energy levels may be represented as a series of lines against a vertical scale of energy, as illustrated in Figure 7.11. The electron in the hydrogen atom can have any of these energy values, but cannot have energies between them.

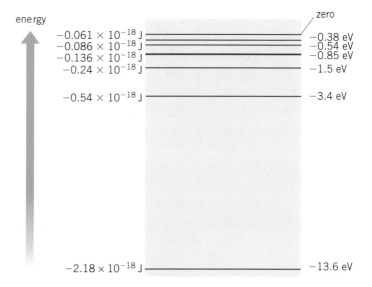

Figure 7.11 Electron energy levels for the hydrogen atom

Normally electrons occupy the lowest energy levels available. Under these conditions the atom and its electrons are said to be in the **ground state**.

Figure 7.12a represents a hydrogen atom with its single electron in the lowest energy state.

If, however, the electron absorbs energy, perhaps by being heated, or by collision with another electron, it may be promoted to a higher energy level. The energy absorbed is exactly equal to the difference in energy of the two levels. Under these conditions the atom is described as being in an **excited state**. This is illustrated in Figure 7.12b.

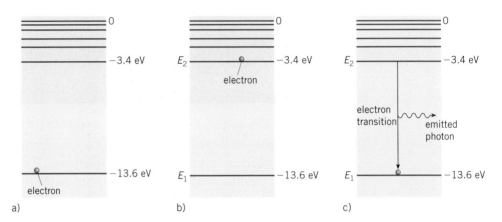

Figure 7.12 Hydrogen atom a) in its ground state, b) in an excited state, c) returning to its ground state with photon emission

An excited atom is unstable. After a short time, the excited electron will return to a lower level. To achieve this, the electron must lose energy. It does so by emitting a photon of electromagnetic radiation, as illustrated in Figure 7.12c. The energy hf of the photon is given by

$$hf = E_2 - E_1$$

where E_2 is the energy of the higher level and E_1 that of the lower, and h is the Planck constant. Using the relation between the speed c of light, frequency f and wavelength λ, the wavelength of the emitted radiation is given by

$$\lambda = \frac{hc}{\Delta E}$$

where $\Delta E = E_2 - E_1$. This movement of an electron between energy levels is called an **electron transition**. Note that the larger the energy of the transition, the higher the frequency (and the shorter the wavelength) of the emitted radiation.

Note that this downward transition results in the **emission** of a photon. The atom can be raised to an excited state by the **absorption** of a photon, but the photon must have just the right energy, corresponding to the difference in energy of the excited state and the initial state. So, a downward transition corresponds to photon emission, and an upward transition to photon absorption.

Spectroscopy

Figure 7.13 shows some of the possible transitions that might take place when electrons in an excited atom return to lower energy levels. Each of the transitions results in the emission of a photon with a particular wavelength. For example, the transition from E_4 to E_1 results in light with the highest frequency and shortest wavelength. On the other hand, the transition from E_4 to E_3 gives the lowest frequency and longest wavelength.

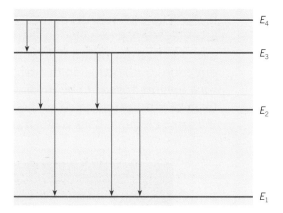

Figure 7.13 Some possible electron transitions

Because all elements have different energy levels, the energy differences are unique to each element. Consequently, each element produces a different and characteristic line spectrum. Spectra can be used to identify the presence of a particular element. The line spectrum of mercury is shown in Figure 7.14. The study of spectra is called **spectroscopy**, and instruments used to measure the wavelengths of spectra are **spectrometers**. Spectrometers for accurate measurement of wavelength make use of diffraction gratings (see Chapter 6) to disperse the light.

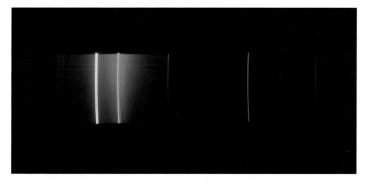

Figure 7.14 Mercury line spectrum

Continuous spectra

Whilst the light emitted by isolated atoms such as those in low-pressure gases produces line spectra, the light emitted by atoms in a solid, a liquid, or a gas at high pressure produces a continuous spectrum. This happens because of the proximity of the atoms to each other. Interaction between the atoms results in a broadening of the electron energy levels. Consequently, transitions of a wide range of magnitudes of energy are possible, and light of a broad spread of wavelengths may be emitted. This is seen as a **continuous spectrum**.

Absorption spectra

If white light passes through a low-pressure gas and the spectrum of the white light is then analysed, it is found that light of certain wavelengths is missing. In their place are dark lines. This type of spectrum is called an **absorption spectrum**; one is shown in Figure 7.15. As the white light passes through the gas, some electrons absorb energy and make transitions to higher energy levels. The wavelengths of the light they absorb correspond exactly to the energies needed to make the particular upward transitions. When these excited electrons return to lower levels, the photons are emitted in all directions, rather than in the original direction of the white light. Thus, some wavelengths appear to be missing. It follows that the wavelengths missing from an absorption spectrum are those present in the emission spectrum of the same element. This is illustrated in Figure 7.16.

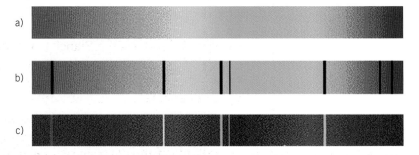

Figure 7.15 An absorption spectrum

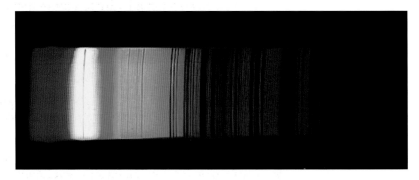

Figure 7.16 Relation between an absorption spectrum and the emission spectrum of the same element: a) spectrum of white light, b) absorption spectrum of element, c) emission spectrum of same element

The line emission or absorption spectrum is a characteristic of the individual element. Such spectra can, therefore, be used to identify elements by comparing the unknown spectrum with one produced under laboratory conditions. For example, the centre of a star consists of gas at very high temperature and pressure, resulting in the emission of white light. This white light passes through the outer cooler layers of the star. What is observed on Earth is an absorption spectrum from the star, enabling identification of the gases in the outer layers.

Example

Calculate the wavelength of the radiation emitted when the electron in a hydrogen atom makes a transition from the energy level at -0.54×10^{-18} J to the level at -2.18×10^{-18} J.
(Planck constant $h = 6.62 \times 10^{-34}$ J s; speed of light $c = 3.00 \times 10^{8}$ m s^{-1})

Here $\Delta E = E_2 - E_1 = -0.54 \times 10^{-18} - (-2.18 \times 10^{-18}) = 1.64 \times 10^{-18}$ J
Using $\lambda = hc/\Delta E$,
$\lambda = 6.62 \times 10^{-34} \times 3.00 \times 10^{8}/1.64 \times 10^{-18} = 1.21 \times 10^{-7}$ m = **121 nm**

Now it's your turn

1 Calculate the wavelength of the radiation emitted when the electron in a hydrogen atom makes a transition from the energy level at -3.4×10^{-18} J to the level at -8.5×10^{-18} J.
(Planck constant $h = 6.6 \times 10^{-34}$ J s; speed of light $c = 3.0 \times 10^{8}$ m s^{-1})

2 The electron in a hydrogen atom makes a transition from the energy level at -13.58 eV to the level at -0.38 eV when a photon is absorbed. Calculate the frequency of the radiation absorbed.
(Planck constant $h = 6.62 \times 10^{-34}$ J s; 1 eV $= 1.60 \times 10^{-19}$ J)

Section 7.3 Summary

- Electrons in atoms can have only certain energies. These energies may be represented in an energy level diagram.
- Electrons in a given energy level may absorb energy and make a transition to a higher energy level.
- Excited electrons may return to a lower level with the emission of a photon. The frequency f of the emitted radiation is given by $E_2 - E_1 = hf$, where E_2 and E_1 are the energies of the upper and lower levels and h is the Planck constant; the wavelength λ is given by $\lambda = c/f$, where c is the speed of light.
- When an electron absorbs energy from white light and moves to a higher energy level, a line absorption spectrum is produced.

Section 7.3 Questions

1 Calculate the wavelength of the radiation emitted when the electron in the hydrogen atom makes a transition from the energy level at -0.54 eV to the level at -3.39 eV.
(Planck constant $h = 6.62 \times 10^{-34}$ J s; speed of light $c = 3.00 \times 10^{8}$ m s^{-1}; 1 eV $= 1.60 \times 10^{-19}$ J)

2 The energy required to completely remove an electron in the ground state from an atom is called the *ionisation energy*. This energy may be supplied by the absorption of a photon, in which case the process is called *photo-ionisation*. Use information from Figure 7.11 to deduce the wavelength of radiation required to achieve photo-ionisation of hydrogen. (Planck constant $h = 6.62 \times 10^{-34}$ J s; speed of light $c = 3.00 \times 10^{8}$ m s^{-1})

Exam-style Questions

1 A zinc plate is placed on the cap of a gold leaf electroscope and charged negatively. The gold leaf is seen to deflect. Explain fully the following observations.
   (a) When the zinc plate is illuminated with red light, the gold leaf remains deflected.
   (b) When the zinc plate is irradiated with ultraviolet radiation, the leaf collapses.
   (c) When the intensity of the ultraviolet radiation is increased, the leaf collapses more quickly.
   (d) If the zinc plate is initially charged positively, the gold leaf remains deflected regardless of the nature of the incident radiation.

2 A beam of monochromatic light of wavelength 630 nm transports energy at the rate of 0.25 mW. Calculate the number of photons passing a given cross-section of the beam each second.
   (Planck constant $h = 6.6 \times 10^{-34}$ J s; speed of light $c = 3.0 \times 10^8$ m s^{-1})

3 The work function energy of a certain metal is 4.0×10^{-19} J.
   (a) Calculate the longest wavelength for which photoemission is obtained.
   (b) This metal is irradiated with ultraviolet radiation of wavelength 250 nm. Calculate, for the emitted electrons:
      (i) the maximum kinetic energy,
      (ii) the maximum speed.
   (Planck constant $h = 6.6 \times 10^{-34}$ J s; speed of light $c = 3.0 \times 10^8$ m s^{-1}; electron mass $m_e = 9.1 \times 10^{-31}$ kg)

4 Calculate the de Broglie wavelengths of:
   (a) a ball of mass 0.30 kg moving at 50 m s^{-1},
   (b) a bullet of mass 50 g moving at 500 m s^{-1},
   (c) an electron of mass 9.1×10^{-31} kg moving at 3.0×10^7 m s^{-1},
   (d) a proton of mass 1.7×10^{-27} kg moving at 3.0×10^6 m s^{-1}.
   (Planck constant $h = 6.6 \times 10^{-34}$ J s)

5 Atoms in the gaseous state (for example, low-pressure gas in a discharge tube) produce an emission spectrum consisting of a series of separate lines. The wavelengths of these lines are characteristic of the particular atoms involved. Hot atoms in the solid state (for example, a hot metal filament in an electric light bulb) produce a continuous emission spectrum which is characteristic of the temperature of the filament rather than of the atoms involved. Suggest reasons for this difference.

6 Figure 7.17 shows some transitions which may take place as excited electrons return to lower energy levels. Calculate the wavelengths of the light emitted in each transition.
   (Planck constant $h = 6.6 \times 10^{-34}$ J s; speed of light $c = 3.0 \times 10^8$ m s^{-1})

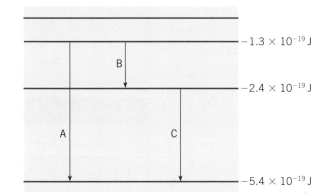

Figure 7.17

7 When the visible spectrum emitted by the Sun is observed closely it is noted that light of certain frequencies is missing and in their place are dark lines.
   (a) Explain how the cool outer gaseous atmosphere of the Sun could be responsible for the absence of these frequencies.
   (b) Suggest how an analysis of this spectrum could be used to determine which gases are present in the Sun's atmosphere.

Answers

Chapter 1

Now it's your turn (page 4)

1 $10\,800\,\text{cm}^2 = 1.08 \times 10^4\,\text{cm}^2$
2 1.0×10^9

Now it's your turn (page 5)

1 kg m^{-3}
2 kg m s^{-2}

Now it's your turn (page 6)

1 **(a)** Yes   **(b)** Yes
2 m^2s^{-2}

Now it's your turn (page 8)

(a) 100–150 g
(b) 50–100 kg
(c) 2–3 m
(d) 0.5–1.0 cm
(e) $0.5\,\text{cm}^3$
(f) $4 \times 10^{-3}\,\text{m}^3$
(g) $220\,\text{m s}^{-1}$
(h) 310 K

Section 1.1 (page 9)

1 **(a)** $6.8 \times 10^{-12}\,\text{F}$ **(b)** $3.2 \times 10^{-5}\,\text{C}$
   (c) $6.0 \times 10^{10}\,\text{W}$
2 800
3 4.6×10^4
4 **(a)** $\text{kg m}^2\text{s}^{-2}$ **(b)** $\text{m}^2\text{s}^{-2}\,\text{K}^{-1}$
5 $\text{kg m}^{-1}\text{s}^{-2}$
6 kg m^2

Now it's your turn (pages 10–11)

1 **(a)** scalar   **(b)** vector   **(c)** scalar
2 **(a)** £32, scalar   **(b)** £8 credit, vector

Now it's your turn (page 13)

1 **(a)** $7\,\text{km h}^{-1}$ **(b)** $1\,\text{km h}^{-1}$
2 11 N at an angle to the 6.0 N force of 56°, in an anticlockwise direction

Now it's your turn (page 15)

1 **(a)** $250\,\text{km h}^{-1}$ **(b)** $180\,\text{km h}^{-1}$
2 **(a)** $1.0\,\text{m s}^{-1}$ **(b)** $9.1\,\text{m s}^{-1}$

Section 1.2 (page 16)

1 **(a)** vector   **(b)** scalar   **(c)** vector
2 Velocity has direction, speed does not. Velocity is speed in a certain direction.
3 Student is correct. Weight is a force which acts vertically downwards.
4 Direction of arrow gives direction of vector. Length of arrow drawn to scale represents magnitude of vector quantity.
5 **(a)** 690 N   **(b)** 210 N   **(c)** 510 N at an angle of 28° to the 450 N force
6 upstream at 78° to the bank
7 120 N at an angle of 25° to the 50 N force in an anticlockwise direction

Exam-style Questions (pages 16–17)

3 **(b) (i)** 92 N   **(ii)** 77 N
   (c) (i) 59 N   **(ii)** 59 N
4 **(a) (i)** $18.3\,\text{m s}^{-1}$ **(ii)** 29° above horizontal
   (b) (i) $10\,\text{m s}^{-1}$ **(ii)** 33°

Chapter 2

Now it's your turn (page 20)

1 $5.3 \times 10^{-11}\,\text{m}$
2 3200 s or 53 min
3 $6\,\text{m s}^{-1}$

Now it's your turn (page 24)

1 $6.0\,\text{m s}^{-2}$
2 3.3 s

Sections 2.1–2.4 (page 25)

1 30 km
2 $180\,\text{m s}^{-1}$
4 3.6 h; $610\,\text{km h}^{-1}$
5 $-5.0\,\text{m s}^{-2}$

Now it's your turn (page 30)

1 $2.5 \, \text{m s}^{-2}$
2 $7.5 \, \text{s}$; $15 \, \text{m s}^{-1}$
3 $8.2 \, \text{m s}^{-1}$ upwards

Now it's your turn (page 32)

(a) $1.8 \, \text{m s}^{-2}$ **(b)** $770 \, \text{m}$

Now it's your turn (page 36)

1 $6.1 \, \text{m s}^{-1}$
2 $20 \, \text{m}$

Sections 2.5–2.7 (page 37)

1 $75 \, \text{m}$
2 $140 \, \text{m s}^{-1}$
3 $47 \, \text{m s}^{-1}$
4 B; A
5 $3.5°$

Now it's your turn (page 41)

1 (i) $0.68 \, \text{s}$;   **(ii)** $6.5 \, \text{m s}^{-2}$
2 The five car-length interval may seem an unsafe under-estimate, but in fact it does make sense. The car in front does not stop instantaneously; once the brake lights go on (implying that the brakes have been applied, and signalling a hazard to the following driver) it will require a distance of $75 \, \text{m}$ (the braking distance) to stop. The following car requires a distance equal to the thinking distance plus the braking distance to stop from 70 m.p.h. The difference is the thinking distance, which is $21 \, \text{m}$ at 70 m.p.h. This is just over five times the average car length of $4 \, \text{m}$.

Exam-style Questions (page 45)

2 8
3 (a) $61°$; $2.8 \, \text{m}$ **(b)** $3.9 \, \text{m s}^{-1}$ **(c)** $1.4 \, \text{s}$
4 (a) $9.4 \, \text{m s}^{-1}$ **(b)** $+11\%$

Chapter 3

Now it's your turn (page 48)

1 (a) $180 \, \text{J}$ **(b)** $30 \, \text{J}$
2 (a) $15.5 \, \text{J}$ **(b)** $7.6 \, \text{J}$

Now it's your turn (page 49)

$1.4 \times 10^4 \, \text{Pa}$

Now it's your turn (page 51)

1 $1800 \, \text{J}$
2 $75 \, \text{J}$

Section 3.1 (page 52)

1 $90 \, \text{J}$; 0; $78 \, \text{J}$; $0.15 \, \text{m}$; $36°$; $2.6 \times 10^3 \, \text{N}$
2 (a) $7.9 \, \text{N}$ **(b)** $0.19 \, \text{J}$
3 $8.66 \times 10^4 \, \text{J}$

Now it's your turn (page 56)

(a) $1.3 \times 10^4 \, \text{J}$ gain   **(b)** $12 \, \text{J}$ gain
(c) $2.4 \times 10^9 \, \text{J}$ loss

Now it's your turn (page 58)

1 $3.1 \times 10^5 \, \text{J}$
2 (a) $1000 \, \text{J}$ **(b)** $3000 \, \text{J}$

Now it's your turn (page 60)

(a) kinetic energy at lowest point → potential energy at highest point → kinetic energy at lowest point, etc.
(b) potential energy of compressed gas → kinetic energy of spray droplets → heat when droplets have stopped moving
(c) kinetic energy when thrown → potential and kinetic energy at highest point of motion → heat and sound energy when clay hits ground

Now it's your turn (page 62)

1 The useful energy output of the heater is in the form of heat energy. All electrical energy is eventually changed to heat energy so the process is 100% efficient.
2 58%

Section 3.2 (page 63)

1 (a) chemical   **(b)** nuclear   **(c)** gravitational potential   **(d)** kinetic   **(e)** sound, light, kinetic energy of gases, potential energy   **(f)** potential
2 (a) $0.51 \, \text{m}$ **(b)** $180 \, \text{J s}^{-1}$
3 (a) $11 \, \text{J}$ **(b)** $14 \, \text{m s}^{-1}$
4 (a) $16 \, \text{MJ}$ **(b)** $1.1 \times 10^2 \, \text{MJ}$ **(c)** $2.1 \times 10^2 \, \text{MJ}$
5 15%

Now it's your turn (page 65)

1 $400 \, \text{N m}^{-1}$
2 (a) $3.0 \times 10^4 \, \text{N m}^{-1}$ **(b)** $54 \, \text{N}$

Now it's your turn (page 66)

1 $5.4 \times 10^{-2} \, \text{J}$
2 $4.2 \, \text{cm}$

Now it's your turn (page 69)

1 **(a)** $1.29 \times 10^7\,Pa$ **(b)** 1.17×10^{-4} **(c)** 0.16 mm
2 $5.3 \times 10^6\,Pa$

Section 3.3 (page 71)

1 **(a)** $500\,N\,m^{-1}$ **(b)** 13.1 cm
2 7.9 J
3 0.36 mm

Now it's your turn (page 73)

1 $4.32 \times 10^5\,J$
2 120 kN
3 1.5 kW

Now it's your turn (page 74)

1 6.7 pence
2 0.96 pence
3 5.3×10^{10}

Section 3.4 (page 75)

1 2100 pence
2 **(a)** **(i)** 16.5 kW **(ii)** 3.2 kN **(iii)** 96 kW
   (b) **(i)** 1.32 MJ **(ii)** 3.84 MJ

Now it's your turn (page 77)

5.6 N m

Now it's your turn (page 78)

4.5 N

Now it's your turn (page 80–81)

1 13.5 N
2 403 N

Section 3.5 (page 83–84)

1 **(a)** 4.2 N m **(b)** 9.1 N
2 **(a)** 2.9 N m **(b)** 8.0 N
3 **(a)** 88 N **(b)** 112 N
4 67 N

Exam-style Questions (page 84)

1 **(c)** **(i)** 1950 J **(ii)** 780 kW

Chapter 4

Now it's your turn (page 89)

1 $1.7\,m\,s^{-2}$
2 5.3 N

Now it's your turn (page 92)

(a) 61 N **(b)** 94 N

Sections 4.1–4.4 (page 93)

1 50 kg
2 **(a)** $7.7\,m\,s^{-1}$ **(b)** $45.2\,m\,s^{-2}$ **(c)** 2040 N
4 $T/2$

Exam-style Questions (page 93)

2 370 N, at 14.2° below horizontal, to left
3 860 N

Chapter 5

Now it's your turn (page 97)

1 2.0 A
2 $4.83 \times 10^5\,s$
3 **(a)** 60 C **(b)** 3.75×10^{20}

Section 5.1 (page 99)

1 760 C
2 $1.9 \times 10^{-4}\,m\,s^{-1}$

Now it's your turn (page 101)

1 $1.6 \times 10^{-13}\,J$
2 **(a)** 0.25 C **(b)** 2.2 J

Now it's your turn (page 103)

100 Ω

Now it's your turn (page 103)

1 current in both is 0.42 A
2 **(a)** 9.2 A **(b)** 26 Ω

Now it's your turn (page 104)

14 minutes

Now it's your turn (page 104)

28.3 pence

Now it's your turn (page 105)

13 A

Section 5.2 (page 106)

1 (a) (i) 0.20 A **(ii)** 0.60 W **(b)** 5400 J
2 3.5×10^6 J
3 2.0 kW
4 £46.32

Now it's your turn (page 110)

1 18 m
2 0.97 mm

Section 5.3 (page 111)

1 6.7 m
2 (b) *I*/A 0.20 0.40 0.60 0.80 1.00 1.20 1.40
   R/Ω 0.95 1.20 2.45 3.65 4.56 5.47 6.21
3 (a) 0.62 Ω **(b)** 4.3×10^{-7} Ω m

Now it's your turn (page 113)

1 0.25 A
2 2.0 A with the current entering the junction

Now it's your turn (page 118)

1 0.56 Ω
2 3.0 A when short circuited; 1.1 W when load resistance equals internal resistance

Section 5.4 (page 119)

1 (a) 0.05 Ω **(b)** 0.3 Ω
2 (a) 0.25 A **(b)** 1.6 Ω **(c)** 12 J

Now it's your turn (page 122)

1 4.3 Ω
2 25 Ω

Now it's your turn (page 125–126)

1 (a) 12 V; **(b)** 0.57 V; **(c)** 5 kΩ
2 (a) 0.8 V; **(b)** 7.7 V

Section 5.5 (page 127)

1 (a) 169 Ω **(b)** 13 Ω
2 (a) 5 Ω **(b)** 3.0 A
3 (a) 25 Ω

Exam-style Questions (pages 127–128)

1 (a) 4.5 V
   (b) (i) 50 Ω **(ii)** 0.090 A **(iii)** 0.90 V
3 ten resistors, each of resistance 12 kΩ and power rating 0.5 W, connected in parallel
4 (a) 4 Ω **(b)** 8 Ω **(c)** 3 Ω **(d)** 1.0 A
5 (a) (i) 1.6×10^{-2} Ω **(ii)** 1.1×10^{-3} Ω
   (iii) 28 W
   (b) (i) 4.4 s **(ii)** 4.4×10^{21}
   (c) (i) 11.7 V **(ii)** 307 W
6 (a) 1.02 V, 1.22 W
   (b) (ii) 7.53 m **(iii)** 1.41 W

Chapter 6

Now it's your turn (page 133)

10

Now it's your turn (page 135)

1 0.40 m s^{-1}
2 0.68 m
3 4.6×10^{14} Hz
4 1.4

Section 6.1 (page 141)

1 (a) 200 Hz **(b)** 5.0×10^{-3} s
2 (a) 7.5×10^{14} Hz to 4.3×10^{14} Hz **(b)** 1.2 m
3 1/4
4 (a) 5.7×10^{-8} J **(b)** 4.9×10^{-5} W m^{-2}

Section 6.2 (page 146)

2 (a) A^2 **(b)** 0 **(c)** $0.25 A^2$

Now it's your turn (page 154)

1 625 nm
2 63 m

Section 6.3 (pages 155–156)

2 (c) 3.5 mm

Now it's your turn (page 161)

1 652 nm
2 28.2°, 70.7°
3 2

Section 6.4 (page 162)

1 (a) 2.5×10^{-6} m **(b)** 10.2° **(c)** 5, 3

Now it's your turn (page 171)

1 880 Hz; 1320 Hz
2 0.38 m
3 340 Hz
4 128 Hz

Section 6.5 (page 172)

2 **(b)** 8.5 cm
3 **(a)** 108 m s^{-1} **(b)** 45 Hz

Exam-style Questions
(page 172–173)

1 2.0×10^9 W m^{-2}
2 **(b)** π rad (180°)   **(c)** 5:1   **(d)** 224 m s^{-1}
4 26.6 cm
6 **(b)** 212 mm   **(c)** 449 mm

Chapter 7

Now it's your turn (page 181)

2 5.8×10^{14} Hz

Section 7.1 (page 182)

1 3.1 eV to 1.8 eV
2 **(a)** 340 nm   **(b)** 2.4×10^{-19} J
3 5.6×10^{-19} J (3.5 eV)

Now it's your turn (page 183)

1 1.3×10^{-11} m
2 1.2×10^{-10} m

Section 7.2 (page 184)

1 3.9×10^{-11} m
2 2.6×10^3 m s^{-1}

Now it's your turn (page 189)

1 3.9×10^{-8} m
2 3.19×10^{15} Hz

Section 7.3 (page 189)

1 430 nm
2 91 nm

Exam-style Questions
(page 190)

2 8.0×10^{14}
3 **(a)** 500 nm
    (b) (i) 3.9×10^{-19} J   **(ii)** 9.3×10^5 m s^{-1}
4 **(a)** 4.4×10^{-35} m   **(b)** 2.6×10^{-35} m
    (c) 2.4×10^{-11} m   **(d)** 1.3×10^{-13} m
6 A: 480 nm; B: 1.8 μm; C: 660 nm

Index

Acknowledgements

The Publishers would like to thank the following for permission to reproduce copyright material:

Picture credits

p.2 Fig. 1.2*r* Rosenfeld Images Ltd/Science Photo Library; Fig. 1.2*l* Royal Observatory Edinburgh/Science Photo Library; Fig. 1.3*l* Andrew Lambert Photography; Fig. 1.3*c* M&C Denis-Hoot/Still Pictures; Fig. 1.3*r* Museum of Flight/Corbis; **p.3** Fig. 1.4 Alfred Pasieka/Science Photo Library; **p.7** Fig. 1.7*b* Hodder Education; Fig. 1.7*t* Yves Lefevre/Still Pictures; **p.10** Fig. 1.9 Action Plus; **p.15** Fig. 1.19 Action Plus; **p.27** Fig. 2.7 Martyn Chillmaid; **p.28** Fig. 2.9 Archivo Iconografico, SA/Corbis; Fig. 2.10 Tony Craddock/Science Photo Library; Fig. 2.11 John McAnulty/Corbis; **p.33** Fig. 2.18 Glyn Kirk/Action Plus; Fig. 2.20 Barry Mayes/Life File; **p.42** Jerome Yeats/Alamy; **p.47** Fig. 3.1 Action Plus; Fig. 3.3 Paul Ridsdale/Alamy; **p.50** Fig. 3.7 Alex Bartel/Science Photo Library; **p.51** Fig. 3.9 Chrysler; **p.53** Fig. 3.11 Andrew Lambert Photography/Science Photo Library; **p.55** Fig. 3.12 Lester Lefkowitz/Corbis; **p.57** Fig. 3.14 Dr Jeremy Burgess/Science Photo Library; **p.60** Fig. 3.15 J-L Charmet/Science Photo Library; **p.74** Fig. 3.26 Clynt Garnham/Alamy; **p.77** Fig. 3.32 Stefan Sollfors/Alamy; **p.87** Fig. 4.2 Bill Sanderson/Science Photo Library; **p.88** Fig. 4.4 Corbis; **p.101** Fig. 5.7 Wendy Brown; **p.107** Fig. 5.12 Science Photo Library; **p.123** Fig. 5.39 Andrew Lambert Photography; **p.144** Fig. 6.20 Polaroid; **p.145** Fig. 6.21 Peter Aprahamian/Sharpless Stress Engineers/Science Photo Library; **p.148** Fig. 6.26 Bruce Coleman Inc./Alamy; **p.152** Fig. 6.35 Andrew Lambert Photography; **p.155** Fig. 6.38 Andrew Lambert Photography; **p.157** Fig. 6.41 Andrew Lambert Photography; **p.162** Fig. 6.48 Bob Winsett/Corbis; **p.164** Fig. 6.52 Andrew Lambert Photography; **p.178** Fig. 7.4 Lucien Aigner/Corbis; **p.183** Fig. 7.7 WA Steer/Daresbury Lab/UCL; Fig. 7.8 Andrew Lambert Photography/Science Photo Library; **p.185** Fig. 7.10 Imperial College/Science Photo Library; **p.187** Fig. 7.14 Imperial College/Science Photo Library; **p.188** Fig. 7.15 Imperial College/Science Photo Library; Fig. 7.16 Mike Guidry, University of Tennessee

Every effort has been made to trace all copyright holders for text and artwork used in this publication. However, if any have been inadvertently overlooked the Publishers will be pleased to make the necessary arrangements at the first opportunity.